Isotopic Assessment of Heterogeneous Catalysis

Isotopic Assessment of Heterogeneous Catalysis

JOHN HAPPEL

Department of Chemical Engineering and Applied Chemistry
Columbia University
New York, New York
and
Catalysis Research Corporation
Palisades Park, New Jersey

 1986

ACADEMIC PRESS, INC.

Harcourt Brace Jovanovich, Publishers
Orlando   San Diego   New York   Austin
Boston   London   Sydney   Tokyo   Toronto

COPYRIGHT © 1986 BY ACADEMIC PRESS, INC.
ALL RIGHTS RESERVED.
NO PART OF THIS PUBLICATION MAY BE REPRODUCED OR
TRANSMITTED IN ANY FORM OR BY ANY MEANS, ELECTRONIC
OR MECHANICAL, INCLUDING PHOTOCOPY, RECORDING, OR
ANY INFORMATION STORAGE AND RETRIEVAL SYSTEM, WITHOUT
PERMISSION IN WRITING FROM THE PUBLISHER.

ACADEMIC PRESS, INC.
Orlando, Florida 32887

United Kingdom Edition published by
ACADEMIC PRESS INC. (LONDON) LTD.
24–28 Oval Road, London NW1 7DX

Library of Congress Cataloging in Publication Data

Happel, John.
  Isotopic assessment of heterogeneous catalysis.

  Bibliography: p.
  Includes index.
  1. Heterogeneous catalysis. 2. Radioisotopes.
I. Title.
QD505.H36 1986   541.3'95   86-7947
ISBN 0–12–324165–0 (alk. paper)

PRINTED IN THE UNITED STATES OF AMERICA

86 87 88 89   9 8 7 6 5 4 3 2 1

To my wife Dorothy

Contents

Preface   xi

Chapter 1 **Introduction**

    1.1 Definitions and Objectives 1
    1.2 An Approach to Reaction Modeling 5
    1.3 Experimental Reactors 11
    1.4 Isotopic Studies of Heterogeneous Catalysis 16
    1.5 Surface Study of Catalysts 20
        References 25

Chapter 2 **Choice of Intermediates and Reaction Steps**

    2.1 Purpose 27
    2.2 Terminal Species and Intermediates 28
    2.3 The Chemisorbed Complexes 28
    2.4 Properties of Catalytic Substances 31
    2.5 Elementary Reaction Steps 35
        References 40

Chapter 3 **Enumeration of Reaction Mechanisms**

    3.1 Independence of Chemical Reactions 41
    3.2 Definitions and Assumptions 44
    3.3 Determination of the General Mechanism 47
    3.4 Listing the Direct Mechanisms 49
    3.5 Systems with a Simple Overall Reaction 52

3.6	Systems with a Multiple Overall Reaction	61
	List of Symbols	72
	References	74

Chapter 4 Superposition Modeling of Mechanisms

4.1	Separation of Mechanism and Rate Equations	76
4.2	The Steady-State Hypothesis	77
4.3	Methods of Tracer Experimentation	78
4.4	General Considerations in Modeling with Tracers	82
4.5	Superposition Modeling with the Gradientless Reactor	86
4.6	Superposition Modeling with the Constant Volume Reactor	92
4.7	Superposition Modeling with the Plug-Flow Reactor	94
	List of Symbols	97
	References	98

Chapter 5 Steady-State Tracing

5.1	Introduction	100
5.2	Step Velocity Grouping	102
5.3	Transition-State Theory and Thermodynamics	105
5.4	Single-Path Reactions	107
5.5	Multiple-Path Reactions	117
	List of Symbols	121
	References	121

Chapter 6 Identifiability and Distinguishability

6.1	Introduction	123
6.2	Identifiability Using Transient Tracing	124
6.3	Distinguishability Using Transient Tracing	129
6.4	General Procedure for Model Testing	133
6.5	Discussion	136
	List of Symbols	136
	References	137

Chapter 7 Transient Tracing

7.1	Introduction	138
7.2	Generalized Data Fitting	140
7.3	Typical Tracer Studies	149
7.4	Tracing with Multiple Marked Species	162
7.5	Conclusions	169
	List of Symbols	170
	References	171

Chapter 8 **Development of Rate Equations**

8.1	Introduction	173
8.2	General Expression for the Kinetics of Simple Reactions	175
8.3	Kinetics of Multiple Reactions	183
8.4	Concluding Remarks	187
	List of Symbols	189
	References	190

Index 193

Preface

In heterogeneous catalysis the combination of reactants with a catalyst produces intermediates that are critical in promoting the formation of desired products. The study of reaction mechanisms is therefore of importance both for the development of effective catalysts and for finding conditions required for their optimal application. Because of the industrial importance of catalysis, a wide variety of approaches has been devoted to achieving a better understanding of how catalysts perform. However, there is still need for a unifying methodology that will serve to coordinate the application of these techniques. The purpose of this book is to show how isotopic tracing can serve to advance this goal.

In the past twenty years a number of advanced spectroscopies for surface observation have yielded valuable information about the nature of adsorbed species present on a catalyst during reaction. Under normal steady-state operation all adsorbed species, including intermediates, are present on the surface at constant concentrations, so it has not been easy to identify those that are indeed involved in a given mechanism.

While kinetic study of overall reaction rates is generally recognized as an essential tool for evaluation of catalyst performance, it is not readily applicable to obtaining fundamental information about reaction mechanisms. However, the substitution of an isotopically marked species into a reaction mixture during an appropriate kinetic experiment without changing operating conditions results in observable isotopic transients that can be modeled in terms of surface concentrations of intermediates and velocities of elementary reaction steps without changing the rate of the overall reaction.

In this book, the use of isotopic tracing is developed in the context of a systematic general approach to chemical reaction mechanisms that could be applied to homogeneous and enzyme catalysis as well as to heterogeneous systems. A new combinatorial method, developed with Peter H. Sellers, deals with the enumeration of all possible reaction mechanisms starting with a given choice of elementary reaction steps. Details of methods of tracer modeling summarize the results of a number of years of research conducted in this field at New York University and Columbia University, and include examples based on experimental study of several industrially important reactions. For computer modeling, a method developed by Jordan L. Spencer is used for generalized data fitting. Based on recent notions in systems analysis, an approach to the study of identifiability and distinguishability of models developed with Eric Walter and Yves Lecourtier is also presented. I am grateful to colleagues in the above studies.

The treatment is aimed at a broad spectrum of readers. Teachers of such subjects as physical chemistry should find it useful in illuminating the study of thermodynamics and kinetics. Research scientists and technologists may be interested in the insight provided for catalyst development and the structure of multiple overall reactions in which by-product formation occurs along with that of the desired product.

I was fortunate in deriving initial motivation for this work from Juro Horiuti, who not only provided seminal ideas but opened doors to contacts with his students and colleagues, some of whom later conducted research with me in the United States. I am grateful to associates at Catalysis Research Corporation for assistance in many ways, especially Miguel A. Hnatow, Laimonis Bajars, and Joseph D. Gettler. Our secretary, Hermine Amirhamzeh, unflinchingly converted illegible manuscript into typescript. Colleagues at the Columbia University Department of Chemical Engineering, especially Huk Y. Cheh, have been most considerate in providing space and encouragement for the conduct of my research in recent years. Juo-Yu Kao devoted much time to reading the manuscript and discussing ideas from the student viewpoint. Finally, my wife Dorothy contributed welcome violin accompaniment to late evening cogitations.

Chapter 1

Introduction

1.1 Definitions and Objectives

In modeling catalytic reactions, the central question concerns the role of the catalyst, which influences the reaction rate of a desired process, and, consequently, the practicality of conducting it. Since the time of Berzelius, who is credited with coining the name *catalyst* for a material apparently not participating in a reaction but yet able to cause it to occur, there has been a tendency to use the term *catalysis* to describe a wide variety of phenomena that are not well understood. Thus, the subject of catalysis is still something of an art as well as a science, but more quantitative theoretical and experimental approaches have become increasingly available.

The definition of Rideal (1968) serves to draw a useful distinction between steps involving intermediates as contrasted to those that only involve terminal species. Catalysis is described as the phenomenon in which one or more substances isothermally (and nonradioactively) augment the rate of a chemical reaction without appearing in the stoichiometric equation of the reaction. What is meant by a stoichiometric equation is a chemical equation that describes the molecules of reactants that are transformed into corresponding molecules of products following the law of conservation of mass. Such an equation does not provide a description of the *mechanism* for the reaction in which individual steps are involved, producing intermediates that remain at constant concentrations during the course of the reaction as the latter proceeds under steady-state conditions. It is the reactions involving these intermediates that are critical in understanding catalytic behavior.

In heterogeneous catalysis the catalyst constitutes one or more phases, which are distinct from reactants and products. In an open system in which the catalyst remains in the form of a fixed bed while reactants enter and products leave, the steady-state concentration of adsorbed intermediates becomes constant.

In dealing with overall or stoichiometric reactions, it is often desirable to consider simultaneous reactions. This is the case, for example, in ethylene oxidation over silver catalyst. It is of interest to know not only the rate of the desired formation of ethylene oxide but also the extent to which secondary formation of carbon dioxide occurs. In more complicated systems involving a number of molecular species, it is important to know just how many independent reactions exist for a given system. Methods for accomplishing this (Aris, 1969) enable us to work with an independent subset of the entire set of reactions involved in a given chemical reaction system. Such a set is satisfactory for simplifying heat and material balance calculations, as well as for facilitating determination of the equilibrium condition reached under a given set of reaction conditions. In the study of mechanisms, however, it is necessary to know detailed reactions among chemisorbed intermediates in addition to reactions involving terminal species.

With reversible processes that approach equilibrium in a reasonable period of time, the presence of a catalyst will not change the ultimate equilibrium concentrations of reactants and products. It is often claimed that in such cases, both forward and backward rates are catalyzed to equal extents (Rideal, 1968). This statement requires some qualification, because catalytic reactions are complex and always consist of sets of mechanistic steps. While the overall appearance and disappearance of molecular species must balance at equilibrium and the exchange rates of corresponding atoms involved must likewise balance, the rates of exchange of the atoms will not always be the same, and, hence, the mechanism will not always be accurately described in terms of single forward and reverse reactions. Although according to the principle of microscopic reversibility, the individual step velocities must also balance at equilibrium they need not be the same for each step. This consideration is important in the development of rate equations for overall reactions by combining velocities of individual steps.

Also if a reaction is studied in the forward direction in the presence of larger concentrations of reactants than products, the concentraction of adsorbed species may be different than when it is studied in the reverse direction at different partial pressures. Consequently, unless a single rate-controlling step prevails close to equilibrium, different mechanistic steps may have low velocities on each side of the equilibrium condition.

Furthermore, the catalyst itself may change composition with changes in reaction conditions. Thus water-gas shift catalysts containing iron will be in a more highly oxidized state at high ratios of H_2O to CO than at lower ratios. Automotive exhaust catalysts are also subject to fluctuations in the oxidation potential of combustion gases, which result in changes of catalyst composition during use.

Even if a catalytic reaction system is studied under constant catalyst composition, it is necessary for the catalyst to interact with ambient substances forming some type of intermediate or intermediates. In some cases the catalyst enters into the reaction as a relatively inert template, providing active sites by chemisorption of reactants, whereas in other instances the catalyst itself participates in the reaction forming intermediate compounds that are generated and consumed in a continuous cycle. A number of instrumental methods have been developed to examine and identify surface structures. A problem with these approaches is that it is often difficult to distinguish between surface species that are involved in a given catalytic reaction and those which though adsorbed do not participate in the reaction. Even with this difficulty, surface examination affords a powerful tool for identification of possible intermediates involved in a mechanistic sequence.

With or without surface examination, it is still necessary to conduct some type of kinetic study to characterize catalyst behavior, since it is the reaction rate that is of ultimate significance in catalyst development and performance. Such experimental data are usually obtained by conducting a reaction under a variety of process conditions in which temperature, pressure, and reactant concentrations are varied. The data obtained consist of observations of overall conversion rates of the reactant species involved, which are correlated by kinetic relationships involving the same variables. Often mathematical rate expressions of some complexity result (Mezaki and Happel, 1969). It is difficult to identify the catalytic mechanism involved, although the resulting expressions may be useful for process design and optimization calculations.

It is therefore desirable to obtain a larger collection of meaningful kinetic data that can be quantitatively used in mechanistic modeling. The use of isotopic tracers affords a unique method for accomplishing this. For each element that participates in a chemical reaction system an atomic tracer can serve to measure the rate of transfer of atoms among reactants and products. The amount of data thus obtained is much larger and enables the choice of viable mechanistic models to be narrowed down more than when only overall rate data are measured.

It is the purpose of this book to present special methods for using isotopic and radioactive atomic species for this purpose. We have studied a number of industrial reactions under steady-state reaction conditions in which superposed tracer transfer was also at *steady state* (Happel, 1972). Use of tracers in this fashion often enables the determination of individual step velocities in an assumed reaction mechanism when the overall reaction is reversible. It is not possible by steady-state tracing to determine concentrations of surface intermediates. We suggested the extension of this technique to studies in which transient-tracer transfer is superposed on a steady-state catalytic reaction system (Happel and Hnatow, 1973) and have demonstrated its applicability in a study of the oxidation of carbon monoxide over Hopcalite catalyst (Happel et al., 1977) and more recently to methanation over nickel and molybdenum sulfide catalysts (Happel et al., 1984). Similar transient techniques have been proposed and used by other investigators.

By using the transient tracer technique, we can estimate surface concentrations of intermediates as well as elementary step velocities. In this procedure, after steady state of an overall catalytic reaction is reached, the feed is switched to one containing exactly the same molecular components but with substitution of one or more marked atomic species for those present in the unmarked feed mixture. Thus the overall reaction remains at steady state but a transient concentration change in the fraction of marked species is observed. The experimental procedure is actually very similar to that employed in steady-state tracer experiments except that we observe tracer redistribution during the period of time required to reach steady state.

In the case of steady-state isotopic transfer, if a gradientless recirculating reactor is employed, the material balances representing isotopic transfer often result in a system of simultaneous linear algebraic equations. In the case of transient tracing, the introduction of the time variable results in a set of simultaneous linear differential equations with constant coefficients. This constitutes a substantial simplification in the interpretation of this type of data.

Since the overall reaction or reactions remain at steady state, catalyst composition and performance should remain constant. In addition, no assumption is necessary with regard to the nature of the kinetics of the individual mechanistic steps. In other techniques in which interpretation of kinetic data is relied on, assumptions about uniformity of sites and interaction between adsorbed species are often necessary, as for example in the case of the familiar Langmuir–Hinshelwood–Hougen–Watson (LHHW) developments. The simplification of treatment afforded by the tracer technique in avoiding problems of reaction kinetics generally makes it a very useful method for study of heterogeneous catalytic systems.

Although the modeling procedure that forms the treatment, which is detailed in later chapters, has not been widely employed using the steady-state tracer technique and hardly at all employing transient tracing, as previously discussed, the use of tracers is not new. Some space in this chapter will be devoted to summarizing isotopic studies in general in order to point out the place of the present technique in heterogeneous catalysis studies.

The methodology developed here might also be considered to be part of the subject referred to variously as "chemical reactor analysis" (Aris, 1969), "chemical reaction engineering," or "applied chemical kinetics" insofar as it is concerned with the development of rate expressions that may be employed in the design and optimization of chemical reactors. Boudart's (1968) book on the kinetics of chemical processes is an excellent introductory approach to the concepts of kinetics from the viewpoint of the chemical engineer. As in that treatment, we limit the scope of subject matter to the chemical reaction rate itself without consideration of the effects of mass and heat transfer, which are important in actual reactor design.

The present treatment is concerned with the use of isotopes to assess velocities of mechanistic steps and concentrations of intermediates in mechanistic models for reaction systems. Such information should be useful in two ways. First, it will enable catalyst development studies to concentrate on the important mechanistic steps of a complex reaction network. Second, it should enable the formulation of more useful rate equations capable of further extrapolation when used in process design. In the next section of this chapter, we expand the concepts of the systematic methodology to be employed in the remaining chapters of the book. Following this, we refer briefly to related subjects which will not be treated further in greater detail, but serve to place the present treatment in context.

1.2 An Approach to Reaction Modeling

In modeling a reaction system two separate problems are encountered. First, it is necessary to know under a given set of conditions, (i.e., fixed ambient gas phase composition, temperature, and pressure) what reactions are occurring. Second, we then wish to determine the effect of changes in phase composition, temperature, and pressure, on the reaction rates. In the case of catalytic reactions that are heterogeneous, it is evident that at least one of the reactants must be adsorbed on the catalyst surface. Usually it is not possible to measure directly under a set of steady-state conditions what rates of intermediate steps are involved. It is possible, however, to maintain constant ambient conditions and to measure the rates of comsumption of terminal species (those supplied as feed to the system or produced as species removed from the system).

A continuous-flow stirred reactor (CSTR) is convenient for this purpose. In such a reaction system a fixed flow of reactants enters the system, which contains a given amount of catalyst. The reactor is designed to secure intimate contact between catalyst and reacting gases, either by placing the catalyst in a basket that is rotated or by rapidly recirculating the ambient gases through the catalyst bed. Thus, the catalyst "sees" the reacting gas mixture at a constant temperature, pressure, and composition of the circulating stream. This gaseous mixture surrounding the catalyst is not the same in composition as the feed that is being continuously introduced at the same time the circulating stream containing products is being withdrawn. A chemical analysis of the difference between inlet and outlet compositions enables the steady-state reaction rate to be estimated. In such a system the external thermal and diffusional gradients around the catalyst particles are minimized. By making the catalyst particles small enough and by operating at a sufficiently low temperature, it is also possible to reduce the internal pore diffusion if the catalyst is in the form of porous pellets or grains.

Chemical analysis enables a gross description of the molecular species to be obtained, from which the number of independent overall chemical reactions corresponding to a given system may be calculated. The problem is then reduced to estimation of the elementary reaction steps, which cannot be measured directly, but which must occur, involving surface species, in order to produce the observed products. If we employ isotopic tracers to study such systems, it is possible to obtain estimates of the "point" condition mechanisms corresponding to a given set of ambient composition, temperature, and pressure without knowledge of how changes in these variables affect the overall rate. This enables the problem of mechanistic modeling to be simplified, because rigorous development of overall reaction rate expressions involves information about the laws governing chemisorption and elementary reaction rates as functions of these variables.

The term *mechanism*, as we will employ it, requires brief examination. Often (Boudart, 1968) it is taken to describe a reaction network or succession of elementary steps that reproduce the stoichiometric equation of a single overall reaction. By elementary steps, we refer to the irreducible acts of reaction in which reactants are transformed into products directly, without passing through an intermediate that is susceptible to isolation or detection. Such elementary steps are considered in classical kinetics, such as transition-state and collision theories. When a system involves more than one such overall reaction, it is only the maximum number of reactions and not the independent reactions themselves that can be identified. Therefore, in this book we have taken the approach that it is more unambiguous to use the term *mechanism* to apply to the reaction network or sequence of steps which applies to a *chemical reaction system* actually studied or observed rather than to the overall chemical reactions among terminal species that can be mathematically deduced from the molecular species produced.

In addition to the velocities corresponding to the elementary reactions involving chemisorbed species, the concentrations of these intermediates also play an important role in characterizing a reaction system. In using transient isotope tracing these concentrations appear as parameters along with the reaction velocities. For a steady-state reaction that consists of a number of unidirectional steps in series, it is not strictly correct to speak of a slow step since, of course, all steps will be occurring at the same rate. However, if the required concentration of intermediates preceding a given step is found to be high as compared with that preceding other steps, it would be inferred that the intrinsic rate constant for that step would be small.

In order to model a reaction system, it is necessary first to assume (1) the intermediate surface species that could enter into the reaction network and (2) the possible elementary steps that might occur involving these species.

Information for selection of surface intermediates may come from a variety of sources. Intermediates present on the catalyst are generally chemisorbed, involving chemical bonds between adsorbate and solid. Consequently, information on rates and amounts of chemisorption of terminal species on the catalyst and whether they are adsorbed molecularly or dissociatively will be useful. It is also sometimes possible by various spectroscopic and other surface science techniques to identify the nature of species present on the catalyst. Another clue comes from the preparation of molecular compounds that will catalyze the reactions involved homogeneously. Rules involving bond energies between atoms may also furnish information about the possible existence of intermediates. This information will, of course, be supplemented by the usual tests of kinetic behavior of the overall reactions.

After selecting a set of surface species, the next step in a systematic approach is the selection of the possible elementary reaction steps. In heterogeneous catalysis an important distinction to be made is whether a step involves a so-called Rideal mechanism in which a gaseous species reacts with an adsorbed species, or a Langmuir mechanism in which two adsorbed species interact with each other. In any case we usually assume that mechanistic steps will consist of simple acts in which a single species reacts to form no more than two products or no more than two species interact with each other.

We are then faced with the combinatorial problem of enumerating the possible ways in which the selected steps and elementary reactions can be combined into viable reaction networks, each of which will yield the observed rates of production of terminal species. Once a set of possible mechanisms has been developed, modeling consists of employing additional data to narrow down the choices.

In the method of analysis considered in this book additional data are provided by experiments in which an isotopic species is introduced into the reaction system. To extend results of tracer kinetics to the traced reaction system, it is necessary to assume that both follow identical input and output pathways. This requires that there be no reaction paths followed exclusively by the tracer such as exchange reactions among species or with the catalyst, unless appropriate corrections are made to allow for such effects. Also effects due to differences in kinetic behavior of the isotopic species from the unmarked atomic species are assumed to be absent.

Each of the combinations of elementary steps and intermediates to be modeled is assumed to be represented by a finite number of discrete interacting states, often called compartments (Berman, 1978). If tracer kinetics follows the traced system, and the reaction system is at steady state, a transient input of tracer into such a system will enable tracer transfer to be represented by a set of simultaneous linear algebraic equations with constant coefficients. These equations will not contain the concentrations of intermediate species, since at steady state the population of both tracer and substance being traced will remain constant.

However, a much richer collection of data is available when the reaction system is maintained steady but the tracer is introduced as a transient in such a fashion that labelled species are substituted for unlabelled. In that case the species present in each compartment must be assumed to be equally accessible. An equal likelihood must exist for both marked and unmarked species to leave a given compartment. For terminal species in the ambient phase, this is readily accomplished by complete mixing, which occurs in a recirculating system. For the chemisorbed species this condition requires that each intermediate be concentrated in a region, all parts of which at steady-state occupancy communicate with each other rapidly so that the mobility of a given chemisorbed species is much higher than the elementary reaction steps in which it participates.

If more than a monolayer is involved in the catalytic reaction, diffusion of external with internal species must be correspondingly rapid compared with rates of elementary steps. If exchange of tracer with sublayers or lattice atoms of the catalyst occurs, it is sometimes possible to allow for such effects by specifying additional compartments with which exchange may occur.

Several additional conditions must be met for convenient interpretation of the transient data. If more than one atom of the type being marked is present in intermediates and terminal species, they must all react at the same rate or information must be available on their rates of reaction. Also the catalyst itself is assumed to remain unchanged for a sufficient period of time to attain a steady-state reaction, or transient behavior of the catalyst will be involved. It is also necessary that the paths of atomic transfer be confined to those that enter into the reaction system being modeled. If other exchange reactions occur, they must be accounted for by separate experiments.

For systems that meet these conditions the mathematical model for following tracer transients assumes the form of a set of linear differential equations with constant coefficients. The unknown parameters appear as the constants in these equations, the independent variable being time, and the dependent variables being the fractional atomic markings of each terminal species and intermediate. Thus, the total number of differential equations is equal to the sum of the number of intermediate and terminal species. The parameters to be determined consist of the concentrations of chemisorbed intermediates together with the reverse velocities of those mechanistic steps that are not unidirectional or at equilibrium. For unidirectional steps, of course, the reverse-step velocity will be zero. In the case of steps at equilibrium, forward and reverse velocities will be equal and very large if they are in the reaction path.

The problem to be solved when we have derived the appropriate equations is not in finding their solution, but the inverse problem of finding the unknown coefficients when given a set of experimental values of tracer transients presumed to be the solution of the differential equations. This is a parameter estimation problem that can be solved by nonlinear regression methods. Details of the procedure (Happel *et al.*, 1977) will be considered in subsequent chapters. The computational procedure that we have employed includes a statistical treatment that provides information for goodness of fit, expected deviation of determined parameters, and degree of correlation among them. In addition to these criteria, the predicted and calculated data should be plotted to observe the possible presence of systematic deviations.

The first problem discussed at the beginning of this section is solved with the determination of reaction velocities and surface concentrations of intermediates corresponding to a satisfactory mechanistic model. This information is directly useful in that it establishes which steps may control the overall rate of reaction.

Modeling of appropriate experimental data will enable determination of reaction-step velocities and surface concentrations of intermediates corresponding to satisfactory mechanistic models. Increased confidence in the validity of such models requires further study of their identifiability and distinguishability. If a study of the structure of a given model reveals that it is not possible in principle to determine the desired parameters, it is clear that it cannot possess fundamental significance. Such a model is said to be unidentifiable. Furthermore, even if the parameters for a model can be identified it may happen that the same data will also enable parameters for a competing model to also be obtained. The models are then said to be indistinguishable. In such cases it may be possible to obtain additional experimental data more closely related to the individual mechanistic steps. These problems are, of course, not unique to tracer studies but are encountered in any type of kinetic modeling.

If a suitable model can be established for a given system, this information can be used to furnish a basis for further catalyst development or construction of rate equations.

In the development of rate equations a basic question at the outset is that of the validity of representing a complex reaction system by means of such equations. The effect of approach to equilibrium on rate in the case of reversible reactions can often be treated in a straightforward fashion. In the case of a reaction in which one step can be considered rate controlling with the other mechanistic steps in quasi-equilibrium, it is possible to separate a "potential" factor from the rate expression that predicts the effect of the thermodynamic equilibrium constant on rate. When more than one step is not at equilibrium the conditions for possible separation of a potential factor can be specified. It is not possible in all cases to employ this convenient method for the prediction of the effect of thermodynamics, although, of course, at equilibrium the overall rate will be zero.

When a thermodynamic potential factor can be separated from the overall rate, the overall reaction can be characterized by unique forward and reverse overall velocities. It is then only necessary to derive an expression for the forward velocity in order to predict the reverse velocity under the same ambient conditions. The accomplishment of this from first principles requires a number of assumptions about the nature of chemisorption and of surface reactions among intermediates.

One of the most popular methods for accomplishing this is the so-called LHHW procedure, developed on the basis of Langmuir's (1918) theory of chemisorption. The mass-action law is employed to predict rates of reaction between adsorbed species or between gaseous reactants and adsorbed species. The basic postulates involved in arriving at rate expressions using this procedure have been questioned, but it appears to lead to useful and reasonable interpretations of experimental data obtained in studies of heterogeneous catalysis.

Efforts to identify the surface concentrations of adsorbed species and rate constants have proven difficult because of the large number of parameters to be estimated, aside from the validity of the underlying assumptions. In some cases the problem has been circumvented by the simple use of power-law expressions to empirically represent forward reaction rates in terms of powers of concentrations or of partial pressures of the reactants involved.

The use of tracers can be helpful here because it furnishes a direct way to assess the importance of the parameters involved in the individual step velocities. It is also possible to directly determine forward and reverse velocities by tracers, a measurement that is impossible using overall kinetic determination of rates.

Even if in principle a reaction can be expressed as the difference between a forward and backward velocity, it may not be possible to express the overall rate analytically in terms of partial pressures of reactants and products. The use of tracers, however, would permit the evaluation of the forward rate at any given degree of conversion of a specified initial feed. A series of such forward rates should be expressible in empirical form, which would be useful for reactor design and control studies.

These two cognate problems of modeling of the mechanism of reaction systems and the development of appropriate rate expressions are treated in further chapters of this book.

1.3 Experimental Reactors

Here we intend to consider apparatus and techniques from the viewpoint of suitability for obtaining the type of measurements and data needed for transient modeling. Recall that the object is to study catalyst performance separate from heat and mass transfer effects. We also assume that the catalyst activity can be maintained at a constant level during the course of experimentation.

A number of reviews are available regarding the design of laboratory and bench-scale reactors. Generally the same basic types of apparatus that are useful for study of catalytic reactions are also used for isotopic tracer studies. Weekman (1974) in discussing the characteristics of various types of laboratory reactors has pointed out the limitations of each, depending on the system studied and information desired. Simpler choices are possible with nondecaying catalysts and where a single reacting phase is involved. Doraiswamy and Tajbl (1974) discussed such systems in some detail. For fundamental studies it is desirable to obtain rate data from isothermal reactors because of the rapid change of reaction rate with temperature. As Bennett *et al.* (1972) have noted, so-called gradientless reactors, in which temperature and concentration gradients are minimized, are especially useful. Before discussing the special problems arising when tracers are employed we will consider a few general features of various laboratory reactors.

The well-mixed batch reactor in which a reaction occurs at constant temperature is often used for studying homogeneous reactions in the liquid phase. For systems involving a gas contacting a solid, fluidized beds with mixing devices have been employed sometimes. A problem in employing such reactors for two-phase systems is that concentrations of components in the gas phase may not parallel those on the catalyst as the batch reaction proceeds. Another disadvantage is that the catalyst is subject to disintegration due to mechanical agitation involved. Therefore, flow reactors containing beds of catalyst particles are more often employed.

The simplest type of bed reactor is one in which the catalyst is contained in a tube through which reacting gases flow. Such flow reactors permit continuous sampling of products and the starting point of the reaction is more easily identified since it occurs at the entrance to the bed. However, changes in concentrations are introduced as the gases pass through the bed. Reactors of this type are usually operated as "integral" converters in which a sufficient degree of reaction occurs so that no problems in chemical analysis are encountered. Radial and longitudinal temperature and concentration gradients may be present in the usual type of integral reactor employed for quick testing of catalysts. By use of tubes of high length-to-diameter ratio and small quantities of catalyst these effects can be minimized, but reaction will still occur over a concentration range.

This can be avoided by operation of the bed as a "differential" reactor in which the rate of conversion is small and may be considered to occur at the arithmetic mean of inlet and outlet concentrations. However, if the total conversion is reduced too much, accurate chemical analysis presents problems.

These difficulties can be resolved by using a recycled reactor, in which the recycle ratio is high enough to achieve an almost uniform composition in the space around the catalyst particles. It is then possible to determine accurately the concentration at which the reaction occurs provided the particles are sufficiently small to avoid intraparticle diffusional and temperature gradients. Such reactors can be operated as either closed or open systems, i.e., with or without net flow of inlet reacting gases. When operated closed the interpretation of data for recycled reactors is the same as for batch reactors. The extent of reactions can sometimes be conveniently followed by the change in total pressure of the system, but the concentration of adsorbed species and the state of the catalyst may vary as the reaction proceeds. It is therefore usually more desirable to operate such systems with a continuous inlet of reactants and removal of the circulating stream of gases so that a steady-state operation is attained. In such a system the catalyst composition and the concentrations of all species in the gas remain constant. It is possible to maintain a substantial degree of conversion by varying the gas feed rate, thus avoiding analytical problems. The effect of reaction products can be determined by introducing them into the feed mixture, but, of course, operation close to initial conditions cannot be achieved without operation as a differential reactor. Thus, such a reactor may be termed a "difference" reactor with internal gradients suppressed. Such reactors have been widely employed in fundamental kinetic studies.

Mathematically, external recirculation is equivalent to any type of completely mixed system in which gaseous feed is continuously introduced and product withdrawn. Both give a direct measure of the rate of reaction from the measured difference between inlet and outlet concentrations. Continuous stirred tank reactors containing fluidized solids have been used in this fashion, but they possess the difficulty of attaining good mixing without catalyst attrition as in the case of stirred batch reactors discussed previously.

Doraiswamy and Tajbl (1974) discuss a number of backmixed reactors that may be used alternatively to the simple externally recycled reactor. In some of these the catalyst is contained in a basket which is attached to a stirrer rotating inside a reaction pot (Carberry, 1964) (Fig. 1.1). A disadvantage of the basket

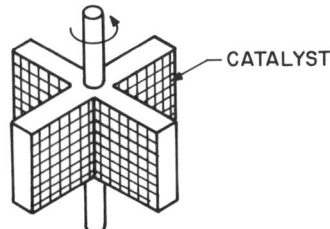

FIG. 1.1 Carberry-type reactor basket.

reactor is that the catalyst surface temperature cannot be accurately measured. There has also been some question as to whether all of the catalyst in such baskets "see" the gas uniformly. In another type of design mixing is achieved by recirculating the reaction mixture internally through a stationary catalyst bed using a specially designed impeller (Berty, 1974) (Fig. 1.2).

Berty (1977) has discussed several new types of recirculating reactors and again emphasized the advantage of gradientless systems for closely attaining the perfectly mixed conditions that are difficult to otherwise achieve. Mahoney et al. (1978) have described variations of the Carberry and Berty reactors used to obtain chemical kinetic data for several important multiphase industrial processes. A new design of all-glass internal recycle reactor has also been developed by Fitzharris and Katzer (1978), and its applicability to catalyst studies has been discussed by Berty (1979), Bartholomew and Erekson (1980), and Fitzharris and Katzer (1980). This variation seems to be especially adapted to studies employing monolithic catalysts. Temkin (1979) has also discussed variations of circulating reactors used by Russian investigators.

More information on kinetics can often be obtained by imposing composition signals on the otherwise steady flow of reactants. Schemes involving integral reactors of this type (Kokes et al., 1955) have been extensively used, but they suffer from difficulty in quantitative interpretation of results of

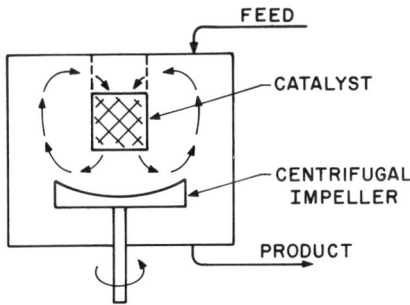

FIG. 1.2 Berty-type reactor.

experiments. A column of catalyst subjected to a pulse exhibits unsteady variations in both time and space, so that partial differential equations are involved in exact analysis. As noted by Bennett *et al.* (1972), the use of pulse and step changes of concentration to a continuous stirred tank reactor involves only time as a variable. In order to secure the highest sensitivity for detection of the amounts of gaseous components adsorbed on the catalyst, it is desirable to keep the ratio of catalyst volume to that of the total reaction system volume as high as possible. This requirement favors the internally recycled reactor when high-pressure operation is involved.

There are some reactor features that are of special applicability to tracer studies. Since traced compounds are costly, it is desirable to operate with relatively small reactors when using stable isotopes that require high concentrations of the tracer elements. Such small reactors present design problems in moving parts such as circulating pumps or turbines and in the means for controlling feed injection.

A number of recirculating pumps have been described for use in external recycling systems. (Doraiswamy and Tajbl, 1974). For some years we employed recirculating reactors of the type originated by Temkin and Horiuti. A typical reactor consisted of a catalyst chamber, a magnetically operated glass piston pump, check valves, and associated connecting tubing. In steady-state tracer studies of the catalytic oxidation of sulfur dioxide (Happel *et al.*, 1971) an improved apparatus of this type was operated isothermally at temperatures in excess of 480°C, which were necessary to prevent condensation in the system. More recently, we have employed a system in which the glass circulating pump was replaced by a stainless steel bellows pump (Metal Bellows Corp., Model MB-41) (Fig. 1.3). For rapid change of feed, two

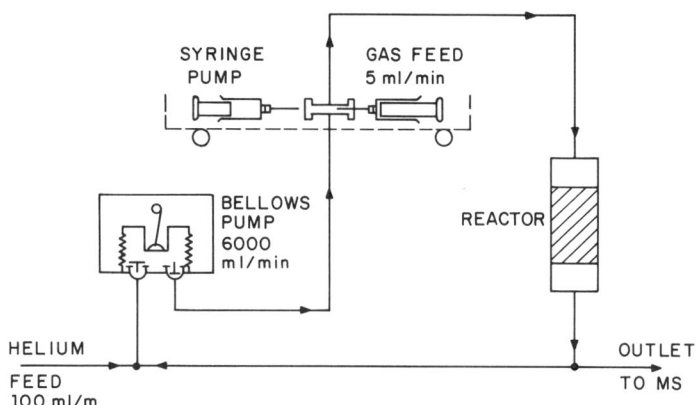

FIG. 1.3 Gradientless recirculating reactor.

1.3 Experimental Reactors

syringe pumps are installed on a movable platform, so that switching to isotopic marking is effected by horizontal motion with a minimum of holdup time in reactor tubing thus providing a step change in marking as discussed in Section 1.1. For operation at elevated reaction temperatures, the gases leaving the reactor are cooled and reheated by a glass exchanger and cooler.

For operating at elevated pressure an all-metal system would be required. Bennett (1972) has described an internally circulated reactor fabricated from metal (Fig. 1.4) and somewhat similar to the Berty-type reactors. In Bennett's reactor a turbine forces gases upward through an annulus containing the catalyst bed and the gas returns downward through a central space. He has devoted special attention to the maintenance of a large catalyst volume to total gas volume so as to maximize the amount of reactant adsorbed compared to that in the dead space.

For analysis of gases containing stable isotopes as tracers various types of mass spectrometers with rapid response times have been employed. Matsumoto and Bennett (1978) reported the analysis of reactants and products from transient experiments of methanation and Fisher–Tropsch reactions using a mass spectrometer with about 12-sec response time (Residual Gas Analyzer 21-614, Consolidated Electrodynamics). Krenzke and Keulks (1980) in transient experiments employing isotopic tracers for propylene oxidation employed a time-of-flight mass spectrometer (Bendix, Model MA-1). The output of the mass spectrometer was digitized by a Columbia Scientific Instruments (CSI-260, digital readout) system, which calculated peak maxima and mass numbers. In recent studies at our laboratories, we have employed a quadrupole mass spectrometer (Finnigan 1015C) equipped with an electronic data processing system.

Although the use of stable isotopes as tracers is generally advantageous because of their ready availability in a number of labelled compounds and the avoidance of safety regulations required in using radioactive isotopes, employment of the latter is sometimes desirable. Sensitivity of detection of

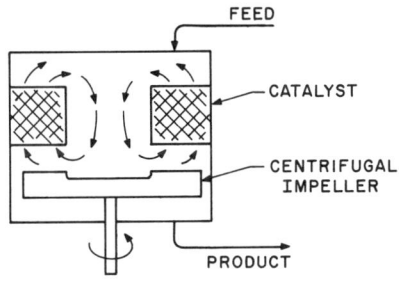

FIG. 1.4 Bennett reactor.

radioactive isotopes is very high through use of scintillation counters (Duncan and Cook, 1968), which are widely employed in medical laboratories. For transient tracing application, requiring rapid analysis for a number of species simultaneously, we have found mass spectrometry to be more useful.

Tracers have been used in a variety of studies of heterogeneous catalysis, some of which have a bearing on the type of applications discussed in this book. A brief review of some of these studies is given in the following section.

1.4 Isotopic Studies of Heterogeneous Catalysis

A monograph by Ozaki (1977) has given a comprehensive review of the various fields of application of isotopes in studies of heterogeneous catalysis. Since the aim of that monograph was not primarily to consider reaction modeling, very little attention was devoted to the techniques discussed in this book. A brief reference is made to the information revealed by replacement of H_2 with D_2 in several reactions, a technique that has features similar to methods we have used. Quantitative interpretation of results of such experiments may be difficult, because of substantial isotopic kinetic effects sometimes introduced by relative differences in mass of the two isotopes. The use of isotopic tracers in studying catalysts and catalytic reactions has also been reviewed by Emmett (1972) and Neiman and Gal (1971).

The use of isotopes in catalysis can generally be divided into several categories in addition to those studied in this book involving modeling of mechanisms. We will briefly consider those that involve (1) studies of catalyst surface and chemisorbed molecules, (2) reaction path studies, and (3) isotope effects, following largely material presented by Ozaki and Emmett.

Studies of the extent of chemisorption of adsorbed molecules are usually made by conventional volumetric or gravimetric techniques. However, the use of radioactive ^{14}C tracers permits the detection of even small amounts of chemisorption. A number of studies have been made, which indicate that very small adsorption values are associated with the catalytic cracking of hydrocarbons. Exchange reactions of hydrocarbons with deuterium have suggested a variety of adsorbed species, and mechanistic implications have been drawn from the results. Hydroxyl groups remaining on an oxide surface after evacuation at high temperatures have been determined by exchange with D_2O and D_2. Oxide coverage on the surface of reduced iron catalysts has been determined by exchange with $H_2^{18}O$. Surface sulfur coverage on metal sulfide catalysts has similarly been determined by exchange with H_2S containing radioactive ^{35}S. Emmett gives further examples of the use of isotopes to measure the extent of adsorption, the nature of adsorbed species, and the degree of coverage of the surface of a catalyst with promoters.

1.4 Isotopic Studies of Heterogeneous Catalysis

Study of the relative rates involved in reaction paths, including transfers in exchange reactions, has perhaps attracted more attention than other applications because of the unique characteristic of tracers to distinguish between alternative paths as well as to determine reverse reaction rates. The use of tracers in this fashion including some of the material to be discussed in this book was the subject of a conference sponsored by the New York Academy of Sciences. (Happel and Hnatow, 1973). Several papers in this conference deal with isotopic exchange reactions.

In exchange reactions no overall chemical reaction occurs. Except for isotopic kinetic effects, the study of such systems is simpler than is the case for reactions in which chemical changes occur. Transfer of atomic species by labeling may be observed as gross transfer from one molecular species to another or, if analytical tools are available, as transfer from given positions in one molecule to those in another, or even as redistribution of a traced atom within a single type of molecular species.

Deuterium has been the most widely employed isotope in heterogeneous catalysis studies, beginning with its isolation in 1932. Burwell (1972) has surveyed the use of deuterium in heterogeneous catalysis, including exchange reactions and other applications. Early studies were conducted for the most part on metals, involving reactions of various hydrocarbons with deuterium. On nickel, for example, it was found that the H—H bond was cleaved most readily. Carbon–hydrogen bonds were cleaved more readily than C—C bonds, and the rate of cleavage of C—H was in the order methane > ethane > propane. More recently there have been many studies of isotopic exchange of alkanes and cycloalkanes on metals aimed at a better understanding of the chemistry of processes that involve dissociative adsorption of a hydrocarbon on a surface. Similar studies have been conducted involving deuterium exchange on a variety of metal oxides (Cr_2O_3, Co_3O_4, NiO, ZnO, Al_2O_3).

Isotopic exchange reactions involving $^{18}O_2/^{16}O_2$ have also been extensively studied and reviewed (Boreskov, 1964). Metal oxides that are active for catalytic oxidation are also active for exchange. Considerable difference is displayed between reaction mechanisms at high and low temperatures. Studies have also been made employing $C^{18}O_2$ in exchange reactions with oxide catalysts.

A number of exchange reactions of other molecules have been conducted. Isotopic equilibration of $^{15}N_2/^{14}N_2$ has been studied in connection with the mechanism of ammonia synthesis. The exchange of H_2O and D_2 over various metals and oxides has also been studied extensively. It is known that OH groups of oxide supports can exchange with D_2O or ND_3 rapidly.

Such exchange reactions are often regarded as test reactions for activation of a particular reactant and thus have served for characterization of catalysts for conducting chemical reactions. Exchange studies are therefore often conducted along with those aimed at the elucidation of the detailed path by which chemisorbed molecules and complexes interact to produce final products.

Reaction path studies, as distinguished from those in which rates of individual elementary steps are determined, form the subject matter of one-third of Ozaki's book. In these studies rates of reverse velocities of individual steps are usually not calculated. A large number of studies of this type have been devoted to various hydrocarbon conversion reactions, which are largely unidirectional. Catalytic cracking finds application in the production of high octane number gasoline, which is rich in branched alkanes and aromatics. Cracking occurs in conjunction with other reactions such as hydrogen transfer, alkylation, cyclization, and polymerization. The nature of intermediate products has been studied extensively by Emmett and co-workers by carrying out cracking in the presence of radioactive ^{14}C compounds and analyzing the products for radioactivity. Various other investigators have studied hydrogenation of olefins by deuterium addition. Ethylene, acetylene, butenes, butadiene, and benzene have been studied in this manner. Rearrangements and isomerizations represent another large class of reactions studied using ^{14}C as tracers. Propylene oxidation to acrolein or reaction in the presence of ammonia to form acrylonitrile has been studied employing ^{14}C and deuterium. The tracer ^{15}N has been employed in study of reactions such as ammonia synthesis.

Attempts to proceed from the identification of reaction paths to a quantitative assessment of reaction rates of the involved elementary steps have been fewer. Several such studies are summarized by Ozaki, including the classic studies of Keii (1953) and Kemball (1956, 1968), using deuterium in olefin hydrogenation; those of Horiuti based on his "stoichiometric number" concept (Horiuti, 1948) to estimate the elementary rates in ammonia synthesis using ^{15}N; and those of Hightower and Hall (1967) in which ^{14}C was employed to determine rate constants in n-butene isomerization reactions. In all these studies the methodology is equivalent to steady-state tracing, a subject covered in later chapters of this book in greater detail, including other reactions such as dehydrogenation, sulfur dioxide oxidation, and the water-gas shift.

Thus far, in this section the applications of isotopes discussed involve those in which isotope substitution does not change the rate of individual mechanistic steps or affect the thermodynamic equilibrium. The main thrust of this book is aimed at developing mechanistic information on this basis, so that the function of the isotope is to act as a tracer. In the case of reactions involving hydrogen the effect of isotopic substitution can be significant. Unfortunately it is often difficult to predict whether a change in rate will result from deuterium substitution. If a change in reaction rate does occur, it is even more difficult to exactly explain the cause in the case of complex reactions such as occur in heterogeneous catalysis. Several books have been written on reaction rates of isotopic molecules. Melander and Saunders (1980) provide an introduction to the subject although the material covered refers almost entirely to homogeneous reactions, which can be conveniently approached by transition-state theory. The final chapter in Ozaki's book gives a review of isotope effects in heterogeneous catalysis and pertinent references for further reading. It must be emphasized that the subject of isotopic effects in heterogeneous catalysis is still largely empirical.

The magnitude of isotopic kinetic effects depends on the location of the isotope in the reacting molecule being studied. If isotopic substitution occurs at the reacting bond, the change in rate is described as a *primary* isotopic effect. Thus C_2H_5OH will react faster than C_2H_5OD, if the splitting of the hydrogen group is rate determining. The effect caused by substitution in more remote bonds is termed *secondary*, as in the case of CD_3 substitution for CH_3 in C_2H_5OH, where the hydrogen group is split. Usually this effect will be smaller than the primary isotope effect, but often complications will arise in heterogeneous catalysis. For example, the rate may be affected by a thermodynamic isotope effect on the concentration of a chemisorbed reaction intermediate rather than the bond splitting involved in the rate-determining step. There is also, of course, no guarantee that transition-state theories can be applied to reactions between chemisorbed species.

Thermodynamic isotope effects may affect the equilibrium constants of the reactions. Calculation of these constants can be made on the basis of established principles of statistical thermodynamics. In the case of isotope exchange reactions, such as reactions of deuterium with hydrocarbons or ammonia, the equilibrium distribution corresponds to randomly deuterated isomers.

Deuterium isotope effects may be determined experimentally by direct rate measurements under steady-state conditions employing the separate protium and deuterium species, respectively. In cases where the isotope is employed as a tracer it is generally not as necessary to employ special procedures except to assume comparable conditions such as reaction temperature, pressure, and activity level of the catalyst. Results can be confirmed by repeated alternations of isotopic species.

Another experimental approach is to arrange the observations as competition between isotopic species in the same reaction mixture. The slower rate of consumption of one isotope can be detected by enrichment by the slower unreacted reactant. If kinetic isotope effects are calculated, it may be desirable to use this method to determine effects due to mixing isotopic species.

Generally, in this book, reactions are modeled in which the isotopic species are used as tracers without correcting for isotopic kinetic effects or making use of conclusions based on such effects. Where such studies provide significant information regarding reaction intermediates or pathways, they are briefly referenced. Generally this is only the case in reactions in which deuterium is employed as a tracer, but such systems have been studied extensively.

Ozaki's comprehensive review of isotopic effects includes such categories as isotope effects in isomerization, on reaction of formic acid and alcohols, on hydrogenations of hydrocarbons and carbon monoxide, on ammonia synthesis and related reactions, and on oxidation reactions. A review of kinetic models in heterogeneous catalysis by Kiperman (1978) devotes considerable attention to the use of different methods of investigation in conjunction with kinetic investigations to establish details of multistage mechanisms. A major portion of his review is devoted to investigations by the author and co-workers in which isotopic kinetic effects were studied. Differences in isotope kinetics between deuterium and protium were analyzed for a variety of reactions including selective hydrogenation of olefins, acetylenic compounds, and aromatic hydrocarbons. Isotope exchange reactions involving hydrocarbons and alcohols were also studied.

The studies reported in this section show that important information regarding reaction intermediates and paths can be obtained by the use of isotopes. In modeling reactions we have found that by marking products it is possible to also obtain useful information about the absence of paths, with resultant simplification in modeling. Another source of information about reaction intermediates is that obtainable by direct surface examination of catalysts using various techniques. A brief summary of this subject is presented in the following section.

1.5 Surface Study of Catalysts

In order to either model a chemical reaction or to study its mechanism, it is necessary to postulate the nature of surface intermediates or chemisorbed species that must necessarily exist in order for a heterogeneous catalyst to operate. Such examination can be conducted by a variety of modern instrumental techniques that observe the catalyst alone or in the presence of reactants. Perhaps the most satisfactory tests of this type are those conducted while a reaction is occurring, preferably during a transient condition so that we have some assurance that the intermediate observed is actually associated with the reaction being studied.

Before considering the applications of surface study specifically to reaction modeling, it is worth noting that the new arsenal of tools and techniques available to surface scientists in recent years has resulted in a considerable surge of effort to understand heterogeneous chemical processes from a fundamental viewpoint. The most successful efforts have been realized in the case of metals and of catalysts based on metal alloys and clusters. Many of the fundamental surface science concepts have been brought together in books by Somorjai (1972, 1981). Sinfelt (1983) has summarized research leading to important industrial applications of catalysts of this type. A review by Yates (1974) is devoted to basic understanding of catalysis using new research techniques for studying kinetics of surface processes, spectroscopy of surface species, and lattice structure of surface layers. Volume II of the series on "Experimental Methods in Catalytic Research" by Anderson and Dawson (1976) is devoted entirely to the characterization of surface and adsorbed species. A book has also appeared which is devoted to the most widely used types of spectroscopy in the field of catalysis (Delgass *et al.*, 1979).

The simplest type of observations of catalyst surfaces involves the use of the optical microscope. As Somorjai (1972) has noted even at a 200× magnification it is possible to observe dislocations on the face of a cadmium sulfide crystal. The electron microscope shows that at 100,000 × magnification, for example on a zinc crystal plane, the surface is full of ledges— small stacks of terraces separated by steps 5–100 Å high. Scientists at Exxon (Baker and Sherwood, 1980) have employed an experimental procedure designated as controlled atmosphere electron microscopy (CAEM). These investigators studied the dynamics of the catalytic oxidation of specimens of graphite on which were deposited fine particles of iridium and rhodium. It was possible to observe the tendency of particles to collect at surface features such as edges or steps during the course of catalysis.

Techniques are also available for viewing surfaces on an atomic level. These include field ion microscopy (FIM) (Mueller and Tsong, 1969) and low-energy electron diffraction (LEED) (Somorjai and Farrell, 1970). In FIM a metallic specimen shaped to a finely pointed needle serves as a cathode while opposite this tip a fluorescent screen acts as an anode, both being mounted in a highly evacuated glass tube. The image information is carried from the tip surface to the screen by radially projected positive ions.

A further development of this technique is the pulsed time-of-flight field ion mass spectrometer (PFITOF) (Beckey, 1971), which can be used for study of surface reactions in ultrahigh vacuum (UHV). Direct kinetic measurements are possible in the millisecond range. The formation of a monolayer at pressures of 10^{-3} torr is completed after 10^{-3} sec and can be desorbed by applying pulsed fields. Parallel experiments of adsorption and dehydrogenation of butanes using tracer techniques and PFITOF, (Thimm *et al.*, 1974) indicated that a firmly bound layer of dehydrogenated species forms the catalytic intermediate on which chemisorption and dehydrogenation to produce butane occur.

Low-energy electron diffraction has been used extensively by Somorjai and co-workers (Somorjai, 1977, 1979) for the study of well-defined surfaces of metals that may serve as catalysts. Chemisorption of adsorbed monolayers of organic molecules and carbon deposits on these surfaces was also studied using this technique. It appears that stepped surfaces show greater reactivity than low-index crystal surfaces in the case of pure metals such as platinum. Somorjai also developed a method for the study of the dynamics of gas–surface interactions, which may be used in conjunction with LEED characterizations of the surface. The method is described as the molecular-beam surface scattering technique. An incident beam of reactants is chopped to provide an unsteady stream which impinges on the catalytic surface. The surface residence time and the reaction velocity are determined by time-of-flight analysis. The technique was applied to study of the exchange reaction between deuterium and hydrogen molecules.

The geometry of the LEED apparatus and auxiliary ultrahigh vaccuum equipment lends itself to the ready application of a family of induced electron emission techniques for surface characterization. Among these may be mentioned Auger electron spectroscopy (AES), X-ray photoelectron spectroscopy (XPS) [also known by the acronym ESCA (electron spectroscopy for chemical analysis)], and ultraviolet induced photoemission spectroscopy (UPS). These three spectroscopies are useful for surface analysis, because most of the information obtained comes from the first few atomic layers. Because of the low-pressure requirement, study under industrial reaction conditions is not possible. Major use appears to be in characterizing a surface before or after a reaction. Auger electron spectroscopy has been applied to surface analysis for the most part. Ultraviolet induced photoemission spectroscopy and XPS have been used to study the chemical state of surface atoms or the nature of interaction of chemisorbed species. Cusumano *et al.* (1978) give representative examples of these applications.

One of the major problems in applying electron spectroscopy techniques is the necessity for operation at ultrahigh vacuum, whereas many catalytic reactions are conducted at substantial pressures. One attempt to overcome this problem is the development by Somorjai (1979) of an apparatus capable of being rapidly pumped down so that it can be alternatively operated at pressures from 10^{-8} to 10^5 torr. In this apparatus the surface area is still quite small (about 1 cm^2) so it is not possible to operate with an appreciable amount of chemisorbed species relative to the dead space holdup when operating at elevated pressures.

Madix (1980) has applied surface science techniques to study catalytic activity using an unsteady temperature technique. In cases that he has studied it appears that surface concentrations of reactants and intermediates can be duplicated at low pressures (10^{-11}–10^{-9} torr) by adjusting the reaction temperature. He and his co-workers have made important contributions to the study of some complicated reactions on metal catalysts using a technique termed temperature-programmed reaction spectroscopy (TPRS).

Temperature-programmed reaction spectroscopy makes use of the principles of temperature programmed desorption in conjunction with spectroscopic techniques. Typically, a desired amount of reactant is dosed to the surface from a gas manifold through a stainless steel needle at a predetermined temperature, and the surface is heated linearly with time. Reaction products then leave the surface at temperatures roughly characteristic of their activation energies and they are detected mass spectrometrically by a quadrupole mass spectrometer. The products do not undergo a significant number of secondary collisions with the surface because of the high pumping speed of the system. The rate–temperature or rate–time data for the formation of each product can be monitored individually.

There are four steps in the interpretation of TPRS. First, the peak temperature of each product is indicative of its rate of formation at the temperature of the peak. Second, it then follows that if a peak temperature for a given product lies above the characteristic desorption temperature for that species (as determined by TPD), a process other than desorption must be rate limiting for its formation. Such observations are clear evidence for surface reaction-limited mechanistic steps. Third, species corresponding to the same peak temperature originate from the same rate-limiting step. Fourth, a quantitative determination of the relative amounts of products desorbed at a common peak temperature provides direct information on the atomic composition of a rate-determining active intermediate. In most cases this intermediate can be identified by analogy to know organometallic complexes or stable compounds. The effects of surface composition and structure can be studied simultaneously to determine their influence on reaction kinetics and mechansims. A number of studies on metal single-crystal surfaces have been conducted for the reactions of carboxylic acids and alcohols that show effects of surface composition and structure on the surface reactions involved. A much clearer picture of surface reactivity is emerging from such model studies.

The pressure limitation does not apply to a wide variety of other spectroscopies in which the measured parameter is the energy associated with a transition between two levels of the same kind. A number of such spectroscopies are discussed by Delgass et al. (1979) including infrared, Raman, Mössbauer, electron spin resonance (ESR), and nuclear magnetic resonance (NMR). Infrared has found wider application to the study of surfaces than any of the others. Raman spectroscopy provides similar information on molecular structure but is much less sensitive. The strong point of Mössbauer spectroscopy is its sensitivity to the chemical environment of the Mössbauer atom. Electron spin resonance is perhaps the most sensitive of these spectroscopies. Under favorable conditions it can be used to determine electronic and geometric structure, symmetry, and reactivity of adsorbed atoms and molecules. The chief application of NMR to surfaces has been to follow relaxation phenomena from which it is possible to deduce molecular motion and magnetic interactions at the surface. Space is not available here to discuss details of these applications, which are adequately discussed in the references cited. We will conclude this section with a brief discussion of the application of infrared spectroscopy simultaneously with transient kinetic studies.

Tamaru (1978) was one of the first to emphasize the importance of techniques suitable for the dynamic treatment of adsorbed species under reaction conditions. As he noted, infrared spectroscopy of adsorbed species is one of those best suited for this purpose. It is possible by using this technique to often detect and identify surface species in catalytic systems under normal reaction conditions. Infrared spectroscopy may be employed to determine the kind and structure of chemisorbed species on the catalyst surface during the reaction at the same time as the rate of the overall reaction is being measured.

Tamaru and his co-workers conducted several studies in which their "dynamic treatment" was applied to catalytic problems. In studies of the decomposition of formic acid over zinc oxide, employing deuterium as a tracer, it was concluded that the rate determining step is the surface decomposition of formate ions. These studies were extended to determine the role of surface formate in the water-gas shift reaction. Further studies were conducted on the decomposition of methyl alcohol on ZnO, and it was found to have much in common with the water-gas shift reaction.

In making such studies Tamaru (1978) suggests the use of a closed circulating system in which initially a known amount of a reacting gas is introduced and circulated through an IR cell containing the catalyst being tested. It is suggested that in some cases a relatively large amount of catalyst also be included in the circulating system so that the amount of adsorption can also be measured volumetrically during the reaction.

A suggested procedure in the case of isotopic tracing is to quickly replace the circulating reactant by a labelled molecule after steady state is achieved. In order to do this in the system described, it would be necessary to rapidly pump out the recirculating mixture and substitute one containing labelled species. Since chemisorption may be a rapid process, such a procedure could result in some perturbation of the steady-state concentration of adsorbed species with consequent difficulty in quantitative interpretation of the data.

The procedure described in Section 1.1 does not present these problems. If the reactor is operated as an *open* rather than *closed* system, it is possible to achieve steady state even at high ratios of catalyst to void volume. When steady state is reached, it is possible by using hardware described in Section 1.3

to switch rapidly from unmarked to marked feed without disturbing the overall steady-state operation.

The combination of various spectroscopies and their use in conjunction with transient techniques will probably receive greater attention in the future. If it were possible to observe transients in concentrations of surface species by improved infrared spectroscopy, this would provide a powerful tool for use in conjunction with transient isotopic tracing.

In this chapter we have attempted to refer the reader to fields of investigation that are pertinent to the subject of modeling by the use of tracers. In addition we have presented an outline of the approach to be followed in subsequent chapters. In the next chapter we intend to begin a more formal development of the subject of modeling in which we explore procedures for selection of possible models.

References

Anderson, R. B., and Dawson, P. T. (1976). "Experimental Methods in Catalytic Research." Academic Press, New York.
Aris, R. (1969). "Elementary Chemical Reactor Analysis". Prentice-Hall, Englewood Cliffs, New Jersey.
Baker, R. T. K., and Sherwood, R. D. (1980). *J. Catal.* **61**, 378.
Bartholomew, C. H., and Erekson, E. J. (1980). *Ind. Eng. Chem. Fundam.* **19**, 132.
Beckey, H. D. (1971). "Field Ion Mass Spectrometry." Pergamon, Oxford.
Bennett, C. O., Cutlip, M. B., and Yang, C. C. (1972). *Chem. Eng. Sci.* **27**, 2255.
Berman, M. (1978). *Prog. Biochem. Pharmacol.* **15**, 92.
Berty, J. M. (1974). *Chem. Eng. Progr.* **70**, 78.
Berty, J. M. (1977). *Prepr. Am. Chem. Soc., Div. Pet. Chem., Chicago, 1977.*
Berty, J. M. (1979). *Ind. Eng. Chem. Fundam.* **18**, 193.
Boudart, M. (1968). "Kinetics of Chemical Processes." Prentice-Hall, Englewood Cliffs, New Jersey.
Boreskov, G. K. (1964). *Adv. Catal.* **15**, 285.
Burwell, R. L., Jr. (1972). *Catal. Rev.* **7**, 25.
Carberry, J. J. (1964). *Ind. Eng. Chem.* **56**, 39.
Cusumano, J. A., Dalla Betta, R. A., and Levy, R. B. (1978). "Catalysis in Coal Conversion." Academic Press, New York.
Delgass, W. N., Haller, G. L., Kellerman, R., and Lunsford, J. H. (1979) "Spectroscopy in Heterogeneous Catalysis." Academic Press, New York.
Doraiswamy, L. K., and Tajbl, D. G. (1974). *Catal. Rev. Sci. Eng.* **10**(2), 177.
Duncan, J. F., and Cook, G. B. (1968). "Isotopes in Chemistry." Oxford Univ. Press (Clarendon), London and New York.
Emmett, P. H. (1972). *Catal. Rev.* **7**, 1.
Fitzharris, W. D., and Katzer, J. R. (1978). *Ind. Eng. Chem. Fundam.* **17**, 130.
Fitzharris, W. D., and Katzer, J. R. (1980). *Ind. Eng. Chem. Fundam.* **19**, 132.
Happel, J. (1972). *Catal. Rev.* **6**(2), 22.
Happel, J., and Hnatow, M. A., eds. (1973). "The Use of Tracers to Study Heterogeneous Catalysis." *Ann. N. Y. Acad. Sci* **213**.
Happel, J., Odanaka, H., and Rosche, P. (1971). *AIChE Sym. Ser.* **67**(115), 60.

Happel, J., Kiang, S., Spencer, J. L., Oki, S., and Hnatow, M. A. (1977). *J. Catal.* **50**, 429.
Happel, J., Cheh, H. Y., Otarod, M., Bajars, L., Hnatow, M. A., and Yin, F. (1984). *Int. Congr. Catal. 8th, Berlin, 1984.*
Hightower, J. W., and Hall, W. K. (1967). *J. Phys. Chem.* **71**, 1014.
Horiuti, J. (1948). *J. Res. Inst. Catal. Hokkaido Univ.* **1**, 8.
Keii, T. (1953). *J. Res. Inst. Catal. Hokkaido Univ.* **3**, 36.
Kemball, C. (1956). *J. Chem. Soc.* 735.
Kemball, C. (1968). *J. Chem. Soc. A* 444.
Kiperman, S. L. (1978). *Russ. Chem. Rev. (Engl. Transl.)* **47**, 3.
Kokes, R. J., Tobin, H., and Emmett, P. H. (1955). *J. Am. Chem. Soc.* **77**, 5860.
Krenzke, L. D., and Keulks, G. W. (1980). *J. Catal.* **61**, 316.
Langmuir, I. (1918). *J. Am. Chem. Soc.* **40**, 136.
Madix, R. J. (1980). *Adv. Catal.* **29**, 1.
Mahoney, J. A., Robinson, K. K., and Myers, E. C. (1978). *Chem. Tech.*, 785, (December 1978).
Matsumoto, H., and Bennett, C. O. (1978). *J. Catal.* **53**, 331.
Melander, L., and Saunders, W. H., Jr. (1980). "Reaction Rates of Isotopic Molecules." Wiley, New York.
Mezaki, R., and Happel, J. (1969). *Catal Rev.* **3**, 241.
Mueller, E. W., and Tsong, T. T. (1969). "Field Ion Microscopy." Elsevier, Amsterdam.
Neiman, M. B., and Gal, D. (1971). "The Kinetic Isotope Method and Its Application." Elsevier, Amsterdam.
Ozaki, A. (1977). "Isotopic Studies of Heterogeneous Catalysis." Academic Press, New York.
Rideal, E. K. (1968). "Concepts in Catalysis." Academic Press, New York.
Sinfelt, J. H. (1983). "Bimetallic Catalysts." Wiley, New York.
Somorjai, G. A. (1972). "Principles of Surface Chemistry." Prentice-Hall, Englewood Cliffs, New Jersey.
Somorjai, G. A. (1977). *Adv. Catal.* **26**, 1.
Somorjai, G. A. (1979). *Surf. Sci.* **89**, 496.
Somorjai, G. A. (1981). "Chemistry in Two Dimensions Surfaces." Cornell Univ. Press, Ithaca, New York.
Somorjai, G. A., and Farrell, H. H. (1970). *Adv. Chem. Phys.* **20**, 215.
Tamaru, K. (1978). "Dynamic Heterogeneous Catalysis." Academic Press, New York.
Temkin, M. I. (1979). *Adv. Catal.* **28**, 173.
Thimm, H., Block, J. H., Strangio, V., and Happel, J. (1974). Joint Meeting of AIChE and VTG, published by AIChE.
Weekman, V. W., Jr. (1974). *AIChE J.* **20**, 833.
Yates, J. T., Jr. (1974). *Chem. Eng. News* Aug. 26, p. 19.

Chapter 2

Choice of Intermediates and Reaction Steps

2.1 Purpose

In modeling by use of tracers it is possible to employ material balances corresponding to production and consumption of species without detailed knowledge of kinetics as a function of changes in partial pressures. Thus, as noted in the previous chapter, it is advantageous to model mechanisms apart from kinetic rate expressions. Before conducting tracer experiments it is desirable to enumerate the mechanisms that should be modeled for a given system. In order to do this it is necessary to select a list of reasonable intermediates and plausible mechanistic steps involving them. This chapter discusses methods for accomplishing this.

In the next chapter a theory based on combinatorial principles will be presented that enables all possible steps to be enumerated once the appropriate terminal species and possible individual mechanistic steps have been selected. A number of examples in heterogeneous catalysis will be worked out in detail. This enumeration is quite general and could be applied in other ways in addition to tracer modeling. With such enumeration, it is possible to proceed with assurance that important possible mechanisms are considered. Once the complete set of possible mechanisms consistent with given assumptions has been enumerated, it is always possible to eliminate some on the basis of further considerations. To interpret tracer data it may be necessary to postulate additional intermediates with the same chemical composition but different activities or species that do not enter into a reaction mechanism but can serve for independent tracer transfer or exchange.

2.2 Terminal Species and Intermediates

The first step in modeling is to specify the terminal products and intermediates. In transient superposition tracing the system to be studied is brought to the steady-state condition in which all components attain constant concentrations before injection of traced species. In heterogeneous catalysis the chemisorbed intermediates that are produced in the course of reaction are assumed to be present at negligible concentrations in the exit stream as far as modeling is concerned. They are produced and consumed at the same rate. Feed and product components present in the effluent stream may, of course, also be chemisorbed.

If a species is consumed at almost the same rate that it is produced but is not completely chemisorbed, it will attain a small steady-state concentration in the effluent stream. If this concentration can be detected by chemical or physical analysis but is still small enough so that it can be neglected in the material balances employed in modeling, it can also be considered to be an intermediate. As the rate of species production begins to exceed that of consumption of a species in the vapor phase, it is necessary to consider it as a by-product. Of course, if a species produced from the feed to a system is not consumed at all, it will be taken as one of the primary products even when present at small concentrations in the effluent.

Intermediates that are adsorbed on the catalyst or exist as solid compounds with it are of prime importance in heterogeneous catalysis. As noted earlier, methods of surface examination can provide useful information about the nature of adsorbing bonds. Infrared spectroscopy has been especially useful in this regard. It must be emphasized, however, that methods of examining the structure of chemisorbed molecules describe only the species to be found there most abundantly whether they participate in the reactions being studied or not.

Thus, the choice of appropriate intermediates is intimately connected with the elementary reaction steps that occur on the catalyst. In advanced stages of modeling of much-studied systems there is often considerable information available to serve as a guide for selecting reaction steps. Often this will not be the case, and it is desirable to proceed in some systematic fashion to develop a set of plausible reaction steps for modeling, including the intermediate species involved.

2.3 The Chemisorbed Complexes

Selection of chemisorbed species requires postulates concerning the nature and the properties of the chemisorbed complexes. In the case of heterogeneous

2.3 The Chemisorbed Complexes

TABLE 2.1. Criteria for Distinguishing between Chemisorption and Physical Adsorption on Surfaces.[a]

Criterion	Chemisorption	Physical adsorption
Heat of adsorption, $-\Delta H_{ads}$	40–800 kJ mole^{-1}	8–20 kJ mole^{-1}
Activation energy, E	Usually small	Zero
Temperature of occurrence	Depends on E, usually low	Depends on boiling point, but usually low
Number of layers adsorbed	Not more than one	More than one possible

[a] Adapted from Bond (1974, p. 16, Table 2.2).

catalysis at least one reactant must be chemisorbed on the surface of the catalyst. Chemisorption must be distinguished from physical adsorption. In the former case there is a rearrangement of electrons within the adsorbed molecule. Physical adsorption on the other hand results from van der Waals forces such as ordinarily lead to liquifaction of gases rather than to the formation of chemical bonds between the adsorbate and the solid catalyst. Table 2.1 (Bond, 1974) summarizes the essential differences between the two types of adsorption.

A further point of distinction between the two types of adsorption concerns their rates. The rate of physical adsorption is always fast, as is the condensation of a vapor on the surface of its own liquid, because there is no activation energy. In practice, it may be difficult to distinguish the two types by their rates because both can occur very rapidly. A slow rate may indicate an activated adsorption, but could also be caused by slow diffusion of physically adsorbed molecules on a porous absorbent. Therefore, great care is needed in interpreting rates of adsorption. Often it is desirable to obtain data over a range of temperatures and pressures.

Physical adsorption generally has little relevance to catalysis, whereas chemisorption is indeed an essential step in the preparation of molecules for reaction. It is beyond the scope of this book to go into details as to how chemisorption occurs and the methods for correlation of chemisorption data. Treatises are available that provide this type of information (Clark, 1970, 1974; Tompkins, 1978). Chapter 3 of Trimm's (1980) book provides a brief review of theories of chemical bonding and adsorption with special reference to catalysis.

Hayward and Trapnell (1964) provide a survey of many chemisorptions of simple gases on metal films, a few of which are listed in Table 2.2. The data in Table 2.2 refer to rates at about 0°C on metal films at low pressures, usually less than 0.001 torr. Much chemisorption under these conditions is very fast; nearly half of the collisions of the gas molecules with the surface result in chemisorption.

TABLE 2.2. Rates of Chemisorption on Metal Films.[a]

Gas	Very fast	Slow	None at 0°C
H_2	Ti, Zr, Nb, Ta, Cr, Mo, W, Fe, Co, Ni, Rh, Pd, Pt, Ba	Mn, Ge	K, Cu, Ag, Cm, Zn, Cd, Al, In, Pb, Sn
O_2	All except Au		As
N_2	La, Ti, Zr, Nb, Ta, Mo, W	Fe, Ba	See H_2; also Ni, Rh, Pd, Pt
CO	See H_2; also La, Mn, Cu	Al	K, Zn, Cds, In, Pb, Sn
CO_2	See H_2; less for Rh, Pd, Pt	Al	Rh, Pd, Pt, Cu, Zn, Cd
CH_4	Ti, Ta, Cr, Mo, W, Rh	Fe, Co, Ni	—
C_2H_4	See H_2; also Cu, Ag	Al	See CO

[a] Adapted from Hayward and Trapnell (1964, Table 2).

Adsorption on oxides differs from that on metals because oxides possess more than one chemical type of adsorption center. In addition, with oxides the term *active surface* need not necessarily mean a surface that contains no adsorbed gas. As a result classification of these chemisorptions is more complicated. A number of isotopic exchange reactions take place at very low temperatures. In many cases, chemisorption at room temperatures is partly rapid and partly slow and sometimes chemisorption seems to be slow even at elevated temperatures. Hayward and Trapnell (1964) provide detailed summaries of available data.

Many simple gaseous molecules are adsorbed with disruption of the bonds within them. For example, in the case of hydrogen adsorption such *dissociative* adsorption may be represented as

$$H_2 + 2l \rightleftharpoons 2Hl \qquad (2.1)$$

where l represents a surface adsorption site, usually on a metal.

Molecules that possess π electrons or lone-pair electrons can chemisorb without such dissociation on two active sites. In the case of ethylene, for example, the following representation may be used

$$C_2H_4 + 2l \rightleftharpoons \underset{\underset{l}{\diagdown}\underset{l}{\diagup}}{H_2C-CH_2} \qquad (2.2)$$

Such molecules are said to be *associatively* chemisorbed. Table 2.3 gives some representative types of adsorbed species involved in hydrocarbon systems. The scheme in Table 2.3 shows a few suggested intermediates for such reactions. The l represents a surface site, and the adsorbed species were named by treating the surface site as a substituent.

In the species shown in Table 2.3 the intermediates in some cases involve the removal of one or two hydrogen atoms from the parent molecule. Thus the monoadsorbed ethane is represented as the half-hydrogenated state C_2H_5-l. Detection of such surface intermediates by infrared spectroscopy has been summarized by Delgass *et al.* (1979).

2.4 Properties of Catalytic Substances

Table 2.3. Surface Intermediates.[a]

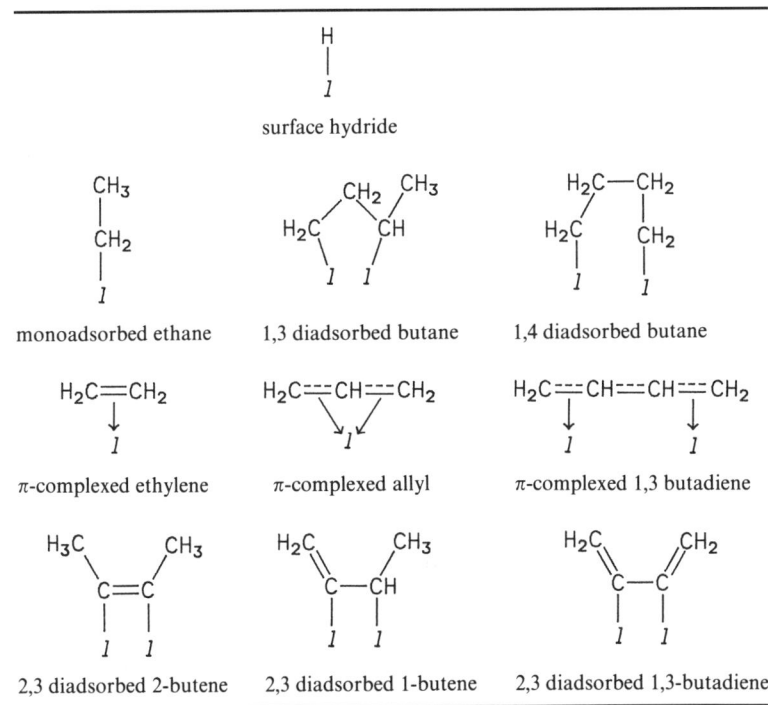

[a] Adapted from Burwell (1973, p. 54, Figure 1).

After primary adsorption of a reactant or reactants additional surface reaction steps can occur on the catalyst. These steps can proceed by either the so-called Langmuir–Hinshelwood mechanisms in which interaction occurs between two chemisorbed species or the Eley–Rideal mechanisms in which molecules from the fluid phase react with a single chemisorbed species. Finally, a step consisting of desorption of one or more products occurs.

2.4 Properties of Catalytic Substances

One of the useful ways to select possible intermediates in the absence of detailed theoretical knowledge is the employment of reaction patterns that correlate the properties of materials with the types of reaction that they catalyze.

A useful distinction that has been observed for many years is the classification of reactions into two types: oxidation–reduction (electronic) or acid–base (ionic). Reactions of the first type include those of oxidation, reduction, hydrogenation, and dehydrogenation. These reactions are catalyzed by solids possessing free or easily excited electrons, i.e., metals or semiconductors. Reactions of the second type include polymerization, isomerization, cracking, dehydration, alkylation, halogenation, and dehalogenation. These reactions are catalyzed by the acidic or basic properties of the catalyst, often proceeded by carbonium ion mechanisms in the case of hydrocarbon reactions.

Texts on catalysis by metals (Bond, 1974) and by oxides and sulfides (Krylov, 1970) are available. Dowden (1973) and Trimm (1980) have summarized general information in this field. Dowden, in setting forth rules for catalyst selection, begins with a consideration of the ambient conditions that correspond to given catalyst compositions.

Such a classification is especially useful for our purposes, since it is assumed that we know the catalyst being employed in the system under study. It is desirable to know the nature of the working catalyst under desired operating conditions, which may alter the catalyst originally charged to a reaction system. For this purpose it is convenient to examine the states assumed by various elements in strongly reducing or oxidizing conditions. A basis for such classification is afforded by the periodic table of elements. Following Dowden's ideas, we will consider the possibilities for metal and compound formation under such conditions.

Table 2.4 presents those elements of interest in heterogeneous catalysis that are capable of existing in stable condition under reducing atmospheres that would be encountered in hydrogenation or dehydrogenation reactions. The conditions chosen are $T \leq 500°C$, $P \sim 1$ atm, the reacting gases being saturated with water vapor at about $25°C$. The elements included are the transition elements. Those in the A subgroups and Group VIII and their alloys, with unfilled d orbitals, have much higher activities than the metallic elements of the B subgroups in this table.

Transition elements in Groups IIIA–VA have oxides that are reduced with great difficulty and, therefore, use in elemental form is improbable. Alkali and

TABLE 2.4. Transition Elements in Reducing Systems.[a]

Period	Group				
	VIA	VIIA	VIII	IB	IIB
4			Fe, Co, Ni	Cu	Zn
5	Mo		Ru, Rh, Pd	Ag	Cd
6	W		Os, Ir, Pt	Au	Hg
7	U	Re			

[a] Adapted from Dowden (1973, p. 642, Table 1).

alkaline earth metals (Groups IA and IIA) can be used under oxygen- and nitrogen-free conditions in some polymerization reactions. The metals Mo, W, V, and Zn have oxides that are difficult to reduce and, therefore, are not used much in elemental form. They would also be sensitive to poisoning by oxygen-containing substances. The elements in Groups IIIB–VIIB are largely nonmetallic and characterized by low melting points. It would therefore be difficult to employ them in the form of high-area heterogeneous catalysts. The elements As, Sb, and Bi, in Group VB and Se and Te in Group VIB possess high vapor pressures, which would result in catalyst loss during operation.

The intrinsic efficiency of metals for hydrocarbon reactions is greatest in Group VIII. Outside Group VIII, finely divided copper has appreciable activity. Thus, in choosing possible intermediates involving chemisorbed species on pure elements, the choice is restricted considerably.

In oxidizing or sulfiding atmospheres the choice of stable elements is even smaller, being confined to the following few precious metals:

$$\begin{array}{ccc} Rh & Pd & Ag \\ Ir & Pt & Au \end{array}$$

In the case of catalysis by metals the chemisorbed reactants and products generally interact only with the catalyst surface. Voorhoeve *et al.* (1976) have termed reactions in such systems as "template" mechanisms. The catalyst may enter into the reaction mechanisms as a relatively inert template providing a collection of atomic orbitals of the proper symmetry and energy. The chemisorption of a number of simple gases has been investigated on many metals (Hayward and Trapnell, 1964).

In catalysis by solid compounds such as oxides and sulfides, it is also useful to refer to the periodic table classification. Table 2.5 (Dowden, 1973) is obtained by eliminating those oxides that are readily reduced. The oxides shown are listed in the most common oxidation states.

TABLE 2.5. Oxides in Reducing Systems.[a]

	Group								
Period	IA	IIA	IIIA	IVA	VA	VIA	VIIIA	IIB	IIIB
2	Li$_2$O*	BeO*							B$_2$O$_3$*
3	Na$_2$O*	MgO*							Al$_2$O$_3$*
4	K$_2$O*	CaO*	Sc$_2$O$_3$*	TiO$_2$	V$_2$O$_5$	Cr$_2$O$_3$	MnO*	ZnO	
5	Rb$_2$O*	SrO*	Y$_2$O$_3$*	ZrO$_2$	Nb$_2$O$_5$	MoO$_3$		CdO	
6	Cs$_2$O*	BaO*	La$_2$O$_3$*	HfO$_2$	Ta$_2$O$_5$	WO$_3$			

[a] Adapted from Dowden (1973, p. 643, Table 3). Asterisk * denotes electronic insulators.

TABLE 2.6. Sulfides in Reducing Systems.[a]

Period	Group						
	IIA	VA	VIA	VIIA	VIII	IB	IIB
4	CaS	V_2S_3	Cr_2S_3	MnS	FeS, CoS, NiS	Cu_2S	ZnS
5			MoS_2			Ag_2S	CdS
6			WS_2	ReS_2			

[a] Adapted from Dowden (1973, p. 644, Table 4).

Silicon dioxide (Group IVB) and P_2O_5 (Group VB) are also insulators. Those oxides in Table 2.5 that are not electronic insulators are semiconductors. As written in their usual oxidation states, they are capable of being reduced to lower oxidation states and, thus, are classified for the most part as n-type semiconductors in reducing atmospheres. By using more extreme reducing conditions they may be converted to a series of still lower oxidation states in some cases. For example in the case of vanadium,

$$V_2O_5 \longrightarrow V_{12}O_{16} \longrightarrow VO_2 \longrightarrow V_2O_3$$

The sulfides of Group VIII are more stable than the oxides. In reducing atmospheres and in the presence of small concentrations of sulfur-containing reactants, the sulfides shown in Table 2.6 may be the stable form expressed in their usual oxidation states.

In the presence of oxygen or water, the oxides listed in Table 2.5 are augmented by the additional oxides given in Table 2.7.

Group IVB, VB, and VIB elements are also generally oxidizable to higher oxidation states.

These tables suggest the importance of ambient conditions for determining the appropriate types of compounds that can exist in the working catalyst,

TABLE 2.7. Additional Oxides in Oxidizing Systems[a,b]

Period	Group				
	IIA	VIIA	VIII	IB	IIB
4		MnO_x	CoO_x, NiO_x	CuO_x	
5				Ag_2O	
6	BaO_x	ReO_x			HgO_x

[a] Adapted from Dowden (1973, p. 644, Table 5).
[b] Subscript x denotes 1–3 depending on severity of oxidizing conditions.

aside from interactions with reacting or intermediate species. In dealing with oxides, sulfides, or similar compounds, it is evident that at least two types of sites for interaction are available, namely those of the metallic cation and those of the anions that are bonded to it. Furthermore, the anions that are present in the catalyst structure can enter into the catalytic processes. For example, in one step of a reaction mechanism, oxygen might be removed from the catalyst by one reagent followed by its replenishment in a subsequent step by another reagent (Voorhoeve et al., 1976).

Insofar as modeling is concerned, it is possible to treat intermediates at steady-state conditions as separate collections of species whether they are involved superficially or intrafacially in a catalytic mechanism. In the latter case, a greater concentration of intermediates may be present than when adsorption and reaction are limited to monolayer coverage.

The nature of the elementary processes that determine the mechanism of a catalytic reaction system is a subject that is at the forefront of present catalysis research. Information being developed in this area can provide an important basis to sharpen the selection of proposed reaction steps. This subject forms the final section of this chapter.

2.5 Elementary Reaction Steps

Once intermediates have been listed, the next step is to develop a set of possible elementary steps involving them. Except for some homogeneous reactions, most chemical reactions are complex, consisting of sets of elementary reactions or steps. A reaction is said to be unimolecular if there is only one molecule in the first member of the chemical equation representing it and bimolecular if there are two. Generally it is assumed unlikely that elementary reactions of molecularity greater than two will occur. Even with this assumption the number of possible elementary steps in most systems becomes very large (see Sinanoglu, 1975; Lee and Sinanoglu, 1981).

One way of reducing the number of possibilities, proposed by Dowden (1973), is to split the overall reactions expected to occur among terminal species into sets of homogeneous reactions, each with a molecularity no greater than two. Intermediates are thus introduced that are also molecular species in the gas phase. These reaction steps are called "virtual" because the actual steps that occur on a catalyst involve chemisorbed species instead of such discrete molecules. A set of steps of this type is termed a "virtual mechanism." Since the equations are all written in terms of specific chemical compounds, it is thus possible to use thermodynamic data to guess which of them is most likely to occur. Such a procedure assumes constancy of the kinetic prefactor that precedes the "potential factor" in reaction rate expressions. Rapid and relatively quantitative estimation of such properties is summarized by Benson (1976).

If the original molecules have any degree of complexity, a description of possibilities using this system can become very involved. Trimm (1980) describes the procedure as applied to practical examples. A number of rules of thumb given by Dowden, including the use of thermodynamic restrictions, are as follows:

(1) the shortest chain is to be preferred, given intermediates of a similar degree of complexity;

(2) chains that develop through relatively stable intermediates are less liable to deviation;

(3) among similar chains those with highly endothermic steps are suspect;

(4) a reactant or intermediate that can decompose by more than one path of a similar kind, as in the elimination of simple molecules, tends to follow the path with the largest decrease in free energy;

(5) the activation energy of an uncoupled, endothermic reaction cannot be less than the enthalpy change.

Since the essence of heterogeneous catalysis lies in the production and reaction of chemisorbed intermediates, it is perhaps preferable to consider the reactions involving these species at the outset rather than first developing "virtual mechanisms;" though rules of thumb, such as those suggested by Dowden, are also useful in the context of mechanisms involving chemisorbed species.

In the remainder of this section, we will review several methods that are applicable to the problem of preliminary consideration of mechanistic steps, although they have not been developed with this as a primary goal.

In earlier studies of catalysis, investigators frequently compared rates of overall reactions with rates of elementary steps determined in separate investigations. Since catalysts differ not only in chemical composition but also in the area of active surface available for reaction corresponding to a unit mass of catalyst, in some cases it becomes customary to report rates referred to surface areas as determined by such procedures as the BET method in which an inert gas such as nitrogen is employed. It has been noted that especially in the case of metallic catalysts, the support may contribute a considerable proportion of the total area measured in this way. The attempt has therefore been made to identify the concentration of "active" centers by adsorption of gases such as hydrogen or carbon monoxide. In the case of metallic catalysts, the number of active sites L has been considered to be close to 10^{15} sites/cm^2, corresponding to an adsorption site for each exposed metal atom. The apparent number of sites for nonmetallic catalysts is usually much smaller. As a further development Boudart (1969) introduced the concept of turnover

2.5 Elementary Reaction Steps

number, corresponding to the number of molecules participating in a reaction per active surface site and per unit time.

Attempts have been made to use the site density for the classification of chemical reactions. Maatman (1980) calculated the concentration of active centers by comparing experimental values of reaction velocity constants to those calculated using formulas based on transition state theory. The determination of the experimental value is assumed to follow the empirical Arrhenius equation

$$k = AC^{-E/RT} \tag{2.3}$$

where A and E are constants. Transition-state theory expressed in terms of thermodynamic formalism is given as approximately

$$k_1 = (kT/h)e^{\Delta S^{\ddagger}/R}e^{-E/RT} \tag{2.4}$$

The universal frequency factor kT/h has the value of 6.34×10^{12} sec^{-1} at $T = 300$ K as developed in transition state theory; $\Delta S^{\ddagger}$ is the entropy change from reactants to the activated complex and can be tabulated using statistical mechanics if the partition functions are known. If k in Eq. (2.3) corresponds to the reaction velocity constant for a measured surface area of catalyst and k_1 refers to a single site, employment of Eq. (2.4) with an estimated $\Delta S^{\ddagger}$ will enable the site density L to be estimated.

Maatman used this approach to obtain estimates for L for many conceivable rate-determining steps for 100 catalyzed reactions reported in the literature. Since the values of A and E in the Arrhenius equation, necessarily determined by measurement of overall reaction rates, are used in conjunction with Eq. (2.4), which refers to a single elementary step, it is assumed that the reactions studied will be characterized by a single rate-determining step. The object of the calculation is thus to determine which of the calculated site densities would correspond to the appropriate rate-determining step. Of course, if calculation reveals that for a certain postulated rate-determining step the value of L is greater than is physically possible, then that step cannot be rate determining. Conversely, if the calculated value of L is so small that the turnover number (number of molecules reacting per site per unit time) is calculated to be larger than the number of molecules that strike the site in a unit time, then this value of L and this step must also be rejected. Maatman concluded that because of (1) the possibility of finding small L values near the lower physical limit and (2) approximations inherent in the method of calculation, a step should not be rejected based on the calculated value of L if the value was in the large range between 10^5 and 10^{17} sites cm^{-2}. Even though this range is large most calculated L values fall outside the range. Thus, to the extent that the L criterion is valid, it enables us to narrow down the choice of a rate-determining step or in our modeling calculations would enable us to reject some postulated reaction steps. Calculations of this type must be used with caution because of the many problems involved in methods of calculation and measurement.

In view of the wide differences observed between calculated and observed reaction rates, when employing methods such as that of Maatman, it is desirable to be able to investigate the kinetics of elementary steps directly. The combined use of the modern tools of surface science has afforded some useful results on reaction kinetics and mechanisms of elementary steps on metal single-crystal surfaces (Madix, 1980). Temperature-programmed reaction spectroscopy, together with appropriate methods of surface characterization, has been applied to several model systems. Transition-state theory has been applied to results obtained for estimation of alternative interactions in adsorbed layers to correlate autocatalytic decomposition of formic acid or nickel. The kinetics of methanol oxidation on silver also show evidence for lateral interactions between adsorbed methoxy groups. A much clearer picture of surface reactivity is emerging from such model studies.

Krylov (1980) has presented a comprehensive review of studies by Maatman and Madix as well as other authors including Russian contributions to knowledge of elementary processes for the determination of the mechanism of catalytic reactions. He notes that in many cases of heterogeneous catalysis, it is not possible to employ reaction velocity constants analogous to simple uni- and bimolecular processes.

Since in a complex reaction system it is not known whether the concept of a single rate-controlling step always applies, it would be desirable to analyze systems without making use of the experimental activition energy, as required in applications of transition-state theory by Maatman. A scheme for doing this has been proposed by Weinberg *et al.* (see Weinberg, 1973; Weinberg and Merrill, 1975). Absolute rate theory is used in conjunction with Johnston's (1966) bond-energy–bond-order (BEBO) method for gas-phase reactions. The BEBO method allows an estimate of relevant activation energies.

Methods of surface science are still largely devoted to the treatment of reaction steps on metal single-crystal surfaces under ultrahigh vacuum so the results are not always directly applicable to actual catalytic reactions involving porous supports and nonmetallic catalysts. If it is possible to start with more data than are available from steady-state kinetic studies, which determine overall reaction velocity constants and activation energies, the techniques of temperature-programmed desorption (TPD) and temperature-programmed reaction (TPR) appear to be especially useful. The basic procedure is similar to surface science techniques but does not require well-defined metal surfaces and high vacuum. An excellent review of TPD and TPR has been presented by Falconer and Schwartz (1983). In a typical TPD experiment on a supported metal catalyst, a small amount of catalyst

(10–200 mg) is contained in a reactor that can be heated. An inert gas, usually helium at atmospheric pressure, flows over the catalyst. A gas is adsorbed on the surface, usually by pulse injections. After excess gas is flushed out, the catalyst is heated to create a linear rise in temperature with time. A detector downstream measures the composition of the effluent stream as a function of temperature. When the inert carrier is replaced by a reactive gas, or when two reacting gases are coadsorbed, the technique is referred to as TPR. Such experiments can provide information on reaction mechanisms unobtainable from steady-state experiments.

In their review Falconer and Schwartz discuss and summarize methods for quantitative reduction of TPD data. They give a comprehensive review of experimental results obtained by a number of investigators for studies on supported metal catalysts. Use of the same type of technique has been reported by He and Ekerdt (1984) for studies of adsorption and reaction of CO and H_2 on ZrO_2. A distinctive temperature pattern was found for multiple TPD and temperature-programmed decomposition (TPDE) products, leading to a proposed mechanism for the interaction of CO/H_2 on the surface of zirconia to produce methane.

A good illustration of how fundamental studies can be effectively used in the development of reaction mechanisms is provided by Rofer-DePoorter (1981), and applied to the Fischer–Tropsch synthesis. The author points out that the two extremes in viewpoint can be used in understanding a reaction mechanism:

(1) the mechanism of a reaction is simply the path that molecules follow in going from reactant to product (Hine, 1962) and

(2) the "mechanism" of a complex reaction is the list of elementary chemical reactions postulated to explain the observed rates and products (Johnston, 1966).

Although the two viewpoints may refer to the same set of facts, each gives a different type of insight. The first viewpoint is commonly employed, and mechanisms are described in which the intermediates are all linked together so that a minimal path is generated. Rofer–DePoorter instead proposed a set of elementary chemical reactions for the Fischer–Tropsch synthesis.

In the modeling procedure used in this book, we propose to first develop all the mechanisms corresponding to the first definition of mechanism previously given, which can be derived from the set of steps corresponding to the second definition. In this chapter, the goal has been to present techniques for the generation of the appropriate sets of elementary steps to employ in such a procedure.

In the following chapter we will assume that by one method or another the choice of suitable terminal species, intermediates and reaction steps, has been made. Even if the true ultimate elementary steps cannot be determined, it is still useful to conduct such modeling. The rate controlling intermediates for unidirectional reactions will be those present at highest concentrations, so that any very short-lived intermediate at low concentration will not affect the rates of critical reaction steps.

The procedure to be followed is to first assemble all possible mechanisms consistent with the choice of species and steps. In later chapters we will develop the modeling procedure to be employed which is aimed at using tracer data to restrict the list of possible mechanisms, preferably to a single predominant one.

References

Benson, S. W. (1976). "Thermochemical Kinetics." 2nd ed, Wiley, New York.
Bond, G. C. (1974). "Heterogeneous Catalysis: Principles and Applications," Oxford Univ. Press (Clarendon), London and New York.
Boudart, M. (1969). *Adv. Catal.* **20**, 153.
Burwell, R. L., Jr. (1973). "Catalysis—Progress in Research," (Fred Basolo and Robert L. Burwell, Jr., eds.). Plenum, New York.
Clark, A. (1970). "The Theory of Adsorption and Catalysis," Academic Press, New York.
Clark, A. (1974). "The Chemisorption Bond." Academic Press, New York.
Delgass, W. N. Haller, G. L., Kellerman, R., and Lunsford, J. (1974). "Spectroscopy in Heterogeneous Catalysis," p. 48, Academic Press, New York.
Dowden, D. A. (1973). *Chim. Ind. (Milan)* **55**, 639.
Falconer, J. L., Schwartz, N., J. A. (1983). *Catal. Rev. Sci. Eng.* **25**, 141.
Hayward, D. O. and Trapnell, B. M. W. (1964). "Chemisorption," 2nd Ed., Butterworth, London.
He, M. Y., and Ekerdt, J. G. (1984). *J. Catal.* **87**, 238.
Hine, J. (1962). "Physical Organic Chemistry." McGraw-Hill, New York.
Johnston, H. S. (1966). "Gas Phase Reaction Rate Theory," p. 26. Ronald Press, New York.
Krylov, O. V. (1970). "Catalysis by Non-Metals: Rules for Catalyst Selection," Academic Press, New York.
Krylov, O.V. (1980). *Kinet. Katal.* **21**, 79.
Lee, L., and Sinanoglu, O. (1981). *Z. Phys. Chem.* **125**, 129.
Maatman, R. W. (1980). *Adv. Catal.* **29**, 97.
Madix, R. J. (1980). *Adv. Catal.* **29**, 1.
Rofer-DePoorter, C. K. (1981). *Chem. Rev.* **81**, 447.
Sinanoglu, O. (1975). *J. Am. Chem. Soc.* **97**, 2309.
Tompkins, F. C. (1978). "Chemisorption of Gases on Metals," Academic Press, New York.
Trimm, D. L. (1980). "Design of Industrial Catalysts," Elsevier, Amsterdam.
Voorhoeve, R. J. H., Remeika, J. P., and Trimble, L. E. (1976). *Ann. N. Y. Acad. Sci.* **272**, 3.
Weinberg, W. H. (1973). *J. Catal.* **28**, 459.
Weinberg, W. H., and Merrill, R. P. (1975). *J. Catal.* **40**, 268.

Chapter 3

Enumeration of Reaction Mechanisms

3.1 Independence of Chemical Reactions

As discussed in the previous chapter, it is usually possible by chemical and physical analytical techniques to determine the character of the terminal species involved. Thus, for example, in the catalytic dehydrogenation of n-butane to butenes it is found that only a very small proportion of isobutene is produced. Consequently in such a case mechanisms and consistent thermodynamic and kinetic relationships can be developed on the basis of the absence of this species.

Once the terminal species consisting of components entering and leaving the reaction system are identified, it is customary to determine the number of overall stoichiometric chemical reactions that can occur in order to generate product molecules from reactants. As discussed in succeeding sections, we shall find it convenient to generalize this concept, we will refer to systems in which more than a single conventional chemical reaction is required as characterized by a multiple overall reaction space. If two chemical reactions, as usually written in chemistry texts, are required, we will describe the system as having an overall reaction space of two dimensions. As will be seen, this type of definition results in a more comprehensive description of the phenomena involved. The following example illustrates the classical approach for determining of the maximum *number* of overall equations required to describe a chemical system.

Example 3.1.1

Consider the selective oxidation of ethylene over a silver catalyst to produce ethylene oxide. In addition to ethylene oxide formation some byproduct formation of carbon dioxide occurs. The terminal species are C_2H_4, C_2H_4O, O_2, CO_2, and H_2O. In the usual chemical nomenclature we could write two chemical equations,

$$2C_2H_4 + O_2 = 2C_2H_4O$$

$$C_2H_4 + 3O_2 = 2CO_2 + 2H_2O$$

This system could be described by an overall reaction space of two dimensions, characterized by the multiple overall reaction

$$\rho(-O_2 - 2C_2H_4 + 2C_2H_4O) + \sigma(-3O_2 - C_2H_4 + 2CO_2 + 2H_2O)$$

where ρ and σ are unrestricted in value.

However, it is also possible to describe this system by a single overall reaction involving all species and the same products, such as

$$7O_2 + 4C_2H_4 = 2C_2H_4O + 4CO_2 + 4H_2O$$

or

$$6O_2 + 7C_2H_4 = 6C_2H_4O + 2CO_2 + 2H_2O$$

These reactions would correspond to selectivities of 50 or 85.7% (i.e., $\frac{6}{7}$) for the conversion of ethylene to ethylene oxide. Experimental data indicate that constant selectivities are not obtained, so it is necessary to revert to the conclusion that two independent overall reactions are involved.

This does not mean that the two reactions first written are the only two that are necessarily involved. Thus, ethylene oxide as well as ethylene could be oxidized to produce carbon dioxide and water:

$$2C_2H_4O + 5O_2 = 4CO_2 + 4H_2O$$

and this reaction could be combined with either of the first two to describe all possible extents of production of species consistent with the atomic balances involved.

Aris and Mah (1963) have presented a concise method for the derivation of Gibbs's rule of stoichiometry, which provides a method to determine the upper limit on the number of independent reactions (or the dimension of the overall reaction space) needed to describe a chemical system. A system is considered in which there are S distinguishable chemical species that partake in R chemical reactions (or in a reaction space of dimension R). The chemical

3.1 Independence of Chemical Reactions

species involved are denoted by $B_1 \ldots B_s$. Each of the species is composed of a set of elements A_n with $n = 1, \ldots, N$.

A matrix may be written in which each row corresponds to one of the chemical species and each column represents the number of times a given element occurs in the corresponding compound. This matrix designated as **A** is called the matrix of atomic coefficients. Thus, if the compounds to be considered are $KClO_3$, KCl, and O_2, the **A** matrix is obtained by writing the formulas in order and then detaching the atomic subscripts

$$\begin{matrix} K & Cl & O_3 \\ K & Cl & \\ & & O_2 \end{matrix} \quad \text{corresponds to} \quad \mathbf{A} = \begin{bmatrix} 1 & 1 & 3 \\ 1 & 1 & 0 \\ 0 & 0 & 2 \end{bmatrix}$$

The rank of the matrix is given by $N - P$, where P is the number of relations between the columns of **A**. Since the first two columns are identical in this case $N - P = 3 - 1 = 2$, the rank of the A matrix is shown to be related to the number of chemical reactions (or the dimension of reaction space) by the formula

$$R \le S - (N - P) = S - N + P \tag{3.1}$$

Thus, for decomposition of potassium chlorate we have $A_1 = K$, $A_2 = Cl$, $A_3 = O$, and $N = 3$; $B_1 = KClO_3$, $B_2 = KCl$, $B_3 = O_2$, and $S = 2$, or

$$R = 3 - 3 + 1 = 1$$

so there is only one independent reaction (the reaction space has dimension 1). The following example illustrates the procedure as applied to the system considered in Example 3.1.1.

Example 3.1.2

Here we have $A_1 = C$, $A_2 = H$, $A_3 = O$, and $N = 3$, $B_1 = C_2H_4$, $B_2 = C_2H_4O$, $B_3 = O_2$, $B_4 = CO_2$, $B_5 = H_2O$. The **A** matrix is then as follows:

$$\begin{matrix} C_2 & H_4 & \\ C_2 & H_4 & O \\ & & O_2 \\ C & & O_2 \\ & H_2 & O \end{matrix} \quad \text{which yields} \quad \mathbf{A} = \begin{bmatrix} 2 & 4 & 0 \\ 2 & 4 & 1 \\ 0 & 0 & 2 \\ 1 & 0 & 2 \\ 0 & 2 & 1 \end{bmatrix}$$

The columns of **A** are not related, so $P = 0$, $S = 5$, and $N = 3$, and $R \le 5 - 3 + 0 = 2$, and the maximum number of independent reactions is two (or the dimension of the overall reaction space is 2.). For more complicated cases, a simple procedure to determine the number of relationships among the columns of the atom-by-species matrix **A** is given in a book by Aris (1969).

Aris and Mah (1963) also described an experimental approach based on matrix analysis to establish the actual number of chemical reactions occurring. The procedure involves observing the extent of changes in the chemical species during the course of time, although time itself does not appear as a variable. They applied their method to the oxidation of ethylene discussed in Examples 3.1.1 and 3.1.2 and found that two independent reactions are indeed required to account for the observed data. This reaction is also considered in Section 3.6.

The determination of the number of independent reactions (dimension of reaction space) is important in thermodynamic calculations to determine the equilibrium composition, which will be attained in a chemical system after a sufficiently long period of time. Often thermodynamic calculations are based on taking the number of independent reactions equal to that given by Eq. 3.1. Björnborn (1977) has reviewed the literature on this subject and shown the usefulness of restricted equilibrium calculations when the kinetics shows that the number of independent reactions is less than that predicted from the atomic matrix. It will be shown in the following development, that if we assume a knowledge of intermediates and elementary steps in addition to terminal species, it is possible not only to enumerate possible reaction mechanisms, but also to specify exactly the dimension of the overall reaction space without the qualifying restriction of an upper bound.

In order to accomplish this we employ a method based on combinatorial principles, (Happel and Sellers, 1982, 1983). It is assumed that all elementary steps can occur in both directions for purposes of complete enumeration. Once this has been accomplished, some can be ruled out or restricted to certain ranges of operating conditions on the basis of thermodynamic or kinetic considerations. It is possible to apply the method readily by following specific examples, so only a brief general description is given here. Further details and proofs are given in the original papers, including additional examples and discussion of related procedures.

3.2 Definitions and Assumptions

A reaction mechanism may be defined as a combination of specific elementary steps that are taken in appropriate proportions to produce the possible degrees of change of terminal species in a reacting system. Usually more elementary steps may be thought possible than those that are required to generate any single mechanism accounting for an observed overall reaction. It is thus necessary to consider how steps may be combined, given an appropriate initial choice of species and steps.

Regardless of how possible intermediates and elementary steps are selected, the procedure given in this chapter presents a method for the unambiguous enumeration of all possible minimal reaction mechanisms that can generate an observed overall consumption and production of terminal species under given reaction conditions. Such minimal mechanisms are defined as the *direct mechanisms* corresponding to a reacting system. A direct mechanism is minimal in the sense that, if one step is omitted from such a mechanism, then there is no mechanism that can be formed by any linear combination of the remaining steps.

In physical chemistry it is generally assumed that a single such predominant direct mechanism is sufficient to characterize a given system. All mechanisms can be classified in terms of direct mechanisms, and it is possible to consider combinations in which two or more direct mechanisms advance simultaneously at independent rates. Such combinations of direct mechanisms may also be termed mechanisms over the allowable range of such combinations without formation of cyclic paths. Sellers (1983) has discussed methods for accomplishing this type of synthesis. The combinations of increasing numbers of direct mechanisms will finally include all possible steps and thus constitute the most general mechanisms consistent with the initial choice of elementary steps.

In modeling isomerization systems involving pseudo-monomolecular kinetics, mechanisms have been employed in which all possible elementary reactions between individual species are assumed to occur (Wei and Prater, 1962), but for most heterogeneous catalytic systems modeling based on such an assumption becomes too complicated. In fact, the further simplifying assumption in Langmuir–Hinshelwood kinetics is usually made that a single rate-controlling step exists in a characteristic direct mechanism. Other steps in the mechanism are assumed to be at equilibrium. In this book, the more general assumption of a rate controlling direct mechanism is taken as the basis for modeling.

The mathematical problem of enumeration of possible direct reaction mechanisms is well defined if it is put in the context of a chemical system that satisfies the following assumptions:

(1) A chemical system consists of species designated as

$$a_1, a_2, \ldots, a_A$$

and mechanistic steps

$$s_1, s_2, \ldots, s_B$$

All we need to know about each step s_i is that it has a unique elementary reaction $\mathbf{R}(s_i)$ associated with it, which may be written in the form:

$$\mathbf{R}(s_i) = \alpha_{i1}a_1 + \alpha_{i2}a_2 + \cdots + \alpha_{iA}a_A \tag{3.2}$$

where some of the α's are positive integers and some are negative; the remaining species that are not involved are taken with α's equal to zero. The chemical equation, by which $\mathbf{R}(s_i)$ would be expressed, is obtained by setting this sum equal to zero and transposing the negative terms to the other side of the equation.

(2) Every step s_i in the chemical system has a set of scalar values associated with it, expressing its possible degrees of advancement. Let us assume that the degree of advancement σ_i of the step s_i can take any real value. Any step s_i together with its degree of advancement σ_i constitutes a vector, which is written $\sigma_i s_i$, and the set of all possible vectors of this form constitutes an S-dimensional vector space. Likewise the family $s_1, \ldots, s_S$ of steps together with their degrees of advancement $\sigma_1, \ldots, \sigma_S$ constitutes a vector, called a *mechanistic vector*, which is written as

$$\sigma_1 s_1 + \sigma_2 s_2 + \cdots + \sigma_S s_S$$

and the set of all vectors of this form constitutes an S-dimensional vector space, called the *mechanism space*.

(3) Just as every step s_i has an elementary reaction $\mathbf{R}(s_i)$ uniquely associated with it, so every mechanistic vector

$$\mathbf{m} = \sigma_1 s_1 + \sigma_2 s_2 + \cdots + \sigma_s S_s$$

has a *reaction vector* $\mathbf{R}(\mathbf{m})$ uniquely associated with it, which is characterized by

$$\mathbf{R}(\mathbf{m}) = \sigma_1 R(s_1) + \sigma_2 R(s_2) + \cdots + \sigma_S R(s_S) \tag{3.3}$$

Consequently a reaction vector $\mathbf{R}(\mathbf{m})$ is a linear combination of species, which is found explicitly by substituting the value of each $\mathbf{R}(s_i)$, as expressed in (1), into Eq. (3.3) and collecting terms. The set of all possible reaction vectors is a vector space, called the *reaction space*, whose dimension is equal to the maximum number of linearly independent elementary reactions. For every reaction vector there is a valid chemical equation, obtained by setting the vector equal to zero and transposing its negative terms.

(4) It is assumed that there is an unambiguous separation of the species of the chemical system into I intermediates $a_1, a_2, \ldots, a_I$ and T terminal species $a_{I+1}, a_{I+2}, \ldots, a_{I+T}$. Hence $A = I + T$. The set of all reaction vectors that involve only terminal species constitutes the *overall reaction space*. Since a

reaction vector is a linear combination of species, we can define this space as consisting of all reaction vectors in which the coefficients of $a_1, a_2, \ldots, a_I$ are zero.

3.3 Determination of the General Mechanism

Under the preceding assumptions the chemical system in question is completely known as soon as the values of the α's are given. These values can be written in the form of an $S \times A$ matrix:

$$\begin{bmatrix} \alpha_{11} & \alpha_{12} & \cdots & \alpha_{1A} \\ \alpha_{21} & \alpha_{22} & \cdots & \alpha_{2A} \\ \vdots & \vdots & & \vdots \\ \alpha_{S1} & \alpha_{S2} & \cdots & \alpha_{SA} \end{bmatrix}$$

This matrix completely determines the linear transformation **R** that takes every mechanistic vector **m** to a reaction vector **R(m)**. Thus, if **m** is written as a row vector $(\sigma_1, \sigma_2, \ldots, \sigma_s)$ and this row vector is multiplied on the right by the matrix, we then obtain the reaction vector, expressed as a row of coefficients:

$$\left(\sum_{i=1}^{S} \sigma_i \alpha_{i1}, \sum_{i=1}^{S} \sigma_i \alpha_{i2}, \ldots, \sum_{i=1}^{S} \sigma_i \alpha_{iA} \right) \tag{3.4}$$

Accordingly, procedures for enumerating mechanisms must ultimately reduce to a manipulation of the α's.

Let us call **R(m)** the *reaction vector produced by the mechanistic vector* **m**. As we have seen, it takes only a simple matrix multiplication to find which reaction vector is produced by a given **m**. Our fundamental problem is the inverse of this, which is to find all the mechanistic vectors that produce a specified reaction vector. In particular, we want those that produce any overall reaction vector **r**. Since **r** only involves terminal species, the mechanistic vectors to be determined are those that recycle all intermediates on them. Incidental to this problem is the determination of every mechanistic vector z that recycles all species in it. Such a mechanistic vector is called a *cycle* and is formally defined by

$$\mathbf{R}(z) = 0 \tag{3.5}$$

A procedure will be first presented for determining the overall reaction space and the space of all mechanistic vectors that produce overall reaction vectors. The overall reaction space will be characterized by a general expression for any element **r** of the space. The expression derived will be of the form

$$\mathbf{r} = \alpha_{I+1}a_1 + \alpha_{I+2}a_2 + \cdots + \alpha_{I+T}a_{I+T} \tag{3.6}$$

accompanied by linear reactions among the coefficients. For simplicity, if there is more than one linearly independent coefficient, we call **r** the *multiple overall reaction of the system*. Likewise, the space of all mechanistic vectors that produce **r** will be characterized by a general expression for any element **m** of the space. The expression will be of the form

$$\mathbf{m} = \sigma_1 s_1 + \sigma_2 s_2 + \cdots + \sigma_S s_S \tag{3.7}$$

accompanied by linear relations among the coefficients. For simplicity we call **m** the *general mechanism for* **r**.

A reaction vector **r**, such as previously discussed, whose coefficients vary such that all possible values of **r** constitute a vector space of dimension R is called a *simple reaction* if $R = 1$ and a multiple reaction if $R > 1$. By definition a space of dimension $R \geq 1$ contains at most R linearly independent vectors, and any such R vectors are said to *generate* the space. Since the set of generators of a space is not uniquely determined, we will consider a multiple reaction as a space of dimension R, rather than as a linear combination of whatever arbitrary R-independent reaction vectors may be chosen to describe it. Thus, to repeat, instead of describing the oxidation of ethylene, as in Example 3.1.1, by two independent overall reactions, we will prefer to refer to it as a system with an overall reaction of a multiplicity of two. This serves to emphasize the fact that the experimental observation of rates of production or consumption of terminal species, under given conditions of temperature, pressure, and concentration, is a uniquely fixed set regardless of how we choose the independent reactions to describe it. We wish to determine the unique mechanisms that correspond to such a set of observations.

We may now, having defined the concepts involved, describe the $S \times A$ matrix $[\alpha_{ij}]$ of reaction coefficients and the supplementary statement as to which columns are coefficients of intermediates as

$$\begin{array}{c} s_1 \\ s_2 \\ \vdots \\ s_S \end{array} \begin{bmatrix} \alpha_{11} & \alpha_{12} & \cdots & \alpha_{1I} & \alpha_{1(I+1)} & \alpha_{1(I+2)} & \cdots & \alpha_{1(I+T)} \\ \alpha_{21} & \alpha_{22} & \cdots & \alpha_{2I} & \alpha_{2(I+1)} & \alpha_{2(I+2)} & \cdots & \alpha_{2(I+T)} \\ \vdots & \vdots & & \vdots & \vdots & \vdots & & \vdots \\ \alpha_{S1} & \alpha_{S2} & \cdots & \alpha_{SI} & \alpha_{S(I+1)} & \alpha_{S(I+2)} & \cdots & \alpha_{S(I+T)} \end{bmatrix}$$

To the left of each row is a symbol for the step where elementary reaction coefficients are in that row. The first I columns contain the coefficients of the intermediates and the remaining T columns contain the coefficients of the terminal species. Since this is merely a shorthand way of writing a family of S equations of the form of Eq. (3.2), we can multiply entire rows, including the s_i at the beginning, by a constant, or we can add entire rows to each other without altering the validity of the equations. Also, we can permute the columns of α's. If a suitable succession of such operations (Birkhoff and MacLane, 1967) is applied to the preceding matrix, it can be converted to the diagonal form illustrated in Fig. 3.1.

Let us now assume that the matrix of a particular chemical system has been reduced to the form shown in Fig. 3.1. Therefore the α's are numbers and each m_i in the left-most column is a linear combination of the form

$$m_i = \gamma_{i1}s_1 + \gamma_{i2}s_2 + \cdots + \gamma_{1S}s_S \tag{3.8}$$

in which the coefficients are numbers. From this we can read off the overall reaction **r** of the system as well as **m**, the general mechanism for **r**. First we have

$$\mathbf{m} = \mu_{H+1}m_{H+1} + \mu_{H+2}m_{H+2} + \cdots + \mu_S m_S \tag{3.9}$$

The coefficients $\mu_{H+1}, \mu_{H+2}, \ldots, \mu_S$ are any real numbers. Next, notice that the terms after $\mu_{H+R}m_{H+R}$, that is

$$\mu_{H+R+1}m_{H+R+1} + \mu_{H+R+2}m_{H+R+2} + \cdots + \mu_S m_S$$

constitute a general expression for any cycle of the system. Therefore, these terms vanish if **R** is applied to them, so that we can evaluate the overall reaction **r** of the system applying **R** to the other terms of **m** as follows:

$$\mathbf{r} = \mu_{H+1}R(m_{H+1}) + \mu_{H+2}R(m_{H+2}) + \cdots + \mu_{H+R}R(m_{H+R}) \tag{3.10}$$

The coefficients $\mu_{H+1}, \mu_{H+2}, \ldots, \mu_{H+R}$ are unrestricted, as before, and each $R(m_{H+1})$ is explicitly known from line $H + i$ of the diagonalized matrix to be

$$\beta_{(H+i)(I+i)}a_{I+i} + \beta_{(H+i)(I+i+1)}a_{I+i+1} + \cdots + \beta_{(H+i)(I+T)}a_{I+T}$$

Therefore, substituting this value into the expression for **r** [Eq. (3.10)], we arrive at an explicit characterization for **r** as a linear combination of the species, where the coefficients are linear expressions depending on the R unrestricted numbers $\mu_{H+1}, \ldots, \mu_{H+R}$.

3.4 Listing the Direct Mechanisms

We wish to construct all possible mechanisms obtainable from Eq. (3.10). Cycle-free mechanisms can be found by systematically deleting steps from the general mechanism equal to the number of cycles. Such a procedure will

	a_1	$\cdots$	a_H	a_{H+1}	$\cdots$	a_I	a_{I+1}	$\cdots$	a_{I+R}	a_{I+R+1}	$\cdots$	a_{I+T}
m_1	β_{11}	$\cdots$	β_{1H}	$\beta_{1(H+1)}$	$\cdots$	β_{1I}	$\beta_{1(I+1)}$	$\cdots$	$\beta_{1(I+R)}$	$\beta_{1(I+R+1)}$	$\cdots$	$\beta_{1(I+T)}$
$\vdots$	$\vdots$		$\vdots$	$\vdots$		$\vdots$	$\vdots$		$\vdots$	$\vdots$		$\vdots$
m_H	0	$\cdots$	β_{HH}	$\beta_{H(H+1)}$	$\cdots$	β_{HI}	$\beta_{H(I+1)}$	$\cdots$	$\beta_{H(I+R)}$	$\beta_{H(I+R+1)}$	$\cdots$	$\beta_{H(I+T)}$
m_{H+1}	0	$\cdots$	0	0	$\cdots$	0	$\beta_{(H+1)(I+1)}$	$\cdots$	$\beta_{(H+1)(I+R)}$	$\beta_{(H+1)(I+R+1)}$	$\cdots$	$\beta_{(H+1)(I+T)}$
$\vdots$	$\vdots$		$\vdots$	$\vdots$		$\vdots$	$\vdots$		$\vdots$	$\vdots$		$\vdots$
m_{H+R}	0	$\cdots$	0	0	$\cdots$	0	0	$\cdots$	$\beta_{(H+R)(I+R)}$	$\beta_{(H+R)(I+R+1)}$	$\cdots$	$\beta_{(H+R)(I+T)}$
m_{H+R+1}	0	$\cdots$	0	0	$\cdots$	0	0	$\cdots$	0	0	$\cdots$	0
$\vdots$	$\vdots$		$\vdots$	$\vdots$		$\vdots$	$\vdots$		$\vdots$	$\vdots$		$\vdots$
m_S	0	$\cdots$	0	0	$\cdots$	0	0	$\cdots$	0	0	$\cdots$	0

FIG. 3.1 The diagonalized matrix. The $H \times H$ submatrix containing $\beta_{1,1} \cdots \beta_{HH}$ and the $R \times R$ submatrix containing $\beta_{(H+1)(I+1)} \cdots \beta_{(H+R)(I+R)}$ have integers on their main diagonals and zeros below.

develop all direct mechanisms (i.e., it includes steps that, if one is removed, contains no paths). If only one or two cycles are involved, this procedure is satisfactory; but even in the case of two cycles a considerable number of trials may be involved. Consequently it is desirable to have a more efficient procedure, as follows.

The generators for the space of all cycles are of the form:

$$\begin{aligned}\mathbf{m}_{H+R+1} &= \gamma_{11}s_1 + \gamma_{12}s_2 + \cdots + \gamma_{1S}s_S \\ \mathbf{m}_{H+R+2} &= \gamma_{21}s_1 + \gamma_{22}s_2 + \cdots + \gamma_{2S}s_S \\ &\vdots \\ \mathbf{m}_S &= \gamma_{C1}s_1 + \gamma_{C2}s_2 + \cdots + \gamma_{CS}s_S\end{aligned} \quad (3.11)$$

where $C = S - H - R$. The γ's in Eq. (3.4.1) can be written as a $C \times S$ matrix. The total number of $C \times C$ submatrices of a $C \times S$ matrix equals the binominal coefficient

$$\binom{S}{C} = \frac{S!}{C!(S-C)!} \quad (3.12)$$

This is an upper bound on the number of direct mechanisms in the system and would correspond to the number of trials required by the systematic deletion previously discussed.

However, it can be shown that from the complete set of all $C \times C$ submatrices only those that are nonsingular (whose determinant is not zero) corresponds to a maximal cycle free subsystem.

The cycle-free subsystem corresponding to each nonsingular $C \times C$ submatrix is found as follows. Start with all steps $s_1, s_2, \ldots, s_S$ and, if the submatrix contains the column

$$\begin{array}{c}\gamma_{1i} \\ \gamma_{2i} \\ \vdots \\ \gamma_{Ci}\end{array}$$

then strike out step s_i. When this has been down with each submatrix column, we are left with $S - C$ steps, and they constitute a maximal cycle-free subsystem.

To find the direct mechanism corresponding to any cycle-free subsystem, including the maximal cycle-free subsystem developed by that procedure, we need the list of C steps that were struck out to form the subsystem. Take the general mechanism for $\mathbf{r}$ given in Eq. (3.10). Each term is a known linear combination of steps, so this expression can be rewritten as a linear combination of the S mechanistic steps with coefficients that are linear functions of the variables $\mu_{H+1}, \mu_{H+2}, \ldots, \mu_S$. In the cycle-free subsystem under consideration, C of these functions are equal to zero, corresponding to those C steps that are absent from the subsystem. This gives C independent equations, which we solve for the C variables $\mu_{H+R+1}, \mu_{H+R+2}, \ldots, \mu_S$. Using these solutions to eliminate these C variables from the general mechanism, we get the unique direct mechanism for r contained in the subsystem under consideration. Its coefficients depend only on the $S - H - C$ or R unknowns $\mu_{H+1}, \mu_{H+2}, \ldots, \mu_{H+R}$, which are the same unknowns on which **r** depends. Thus, each direct mechanism spans a space of the same dimension as the space spanned by its overall reaction **r**. All direct mechanisms for **r** are obtained by this procedure. There can be repetitions among them, because of the fact that two subsystems can contain the same mechanism for **r**.

In the last two sections of this chapter we will illustrate the procedure developed in this section by a number of examples of increasing complexity. These examples do not in all cases include latest research on the chemistry involved, since we wish to illustrate the method in comparison with similar cases that have been given in the literature. Validation of the mechanisms involved depends on experimental data, which in many cases is not available in sufficient detail. Our purpose in presenting the examples is primarily to illustrate the methodology. We emphasize that the working of these examples can be rather easily followed even if one does not wish to take the time to fully understand the principles and procedures previously discussed.

3.5 Systems with a Simple Overall Reaction

Each chemical system considered in this section has an overall reaction space of dimension 1. In the chemical engineering literature beginning with the pioneering work of Hougen and Watson (1947) it has been customary to consider systems of this type. Furthermore, the mechanisms to be evaluated are generally developed by assuming that each intermediate is produced by a single mechanistic step and consumed by another single step. For such a selection, the sum of the independent intermediates will be just one less than the number of steps in the mechanism and there will be no cycles. The first example follows this simplification.

Note in this and following examples that the process of matrix diagonalization is facilitated by first selecting steps that only involve a single intermediate and numbering the intermediates in corresponding columns. Each succeeding row should only contain intermediates that can be eliminated by employing multiples of previous rows.

3.5 Systems with a Simple Overall Reaction

Example 3.5.1

The oxidation of sulfur dioxide has been modeled by various mechanisms. One that is often employed will serve to illustrate our procedure although, of course, such formalism would not be required for this simple case.

The overall reaction of the system is given as

$$2SO_2 + O_2 = 2SO_3$$

and four mechanistic steps are assumed, whose elementary reactions are given as

$$SO_2 + l \rightleftarrows SO_2 l$$
$$O_2 + 2l \rightleftarrows 2Ol$$
$$SO_2 l + Ol \rightleftarrows SO_3 l + l$$
$$SO_3 l \rightleftarrows SO_3 + l$$

The intermediates are $SO_2 l$, Ol, and $SO_3 l$; l itself need not be considered since it does not involve any terminal species. We characterize the system as

$$R(s_1) = -SO_2 + SO_2 l$$
$$R(s_2) = -O_2 + 2Ol$$
$$R(s_3) = -SO_2 l - Ol + SO_3 l$$
$$R(s_4) = -SO_3 l + SO_3$$

The matrix corresponding to these equations is

$$\begin{array}{c c} & \begin{array}{cccccc} SO_2 l & Ol & SO_3 l & SO_2 & O_2 & SO_3 \end{array} \\ \begin{array}{c} s_1 \\ s_2 \\ s_3 \\ s_4 \end{array} & \left[\begin{array}{cccccc} 1 & 0 & 0 & -1 & 0 & 0 \\ 0 & 2 & 0 & 0 & -1 & 0 \\ -1 & -1 & 1 & 0 & 0 & 0 \\ 0 & 0 & -1 & 0 & 0 & 1 \end{array} \right] \end{array}$$

Diagonalization gives

$$\begin{array}{c c} & \begin{array}{cccccc} SO_2 l & Ol & SO_3 l & SO_2 & O_2 & SO_3 \end{array} \\ \begin{array}{c} s_1 \\ s_2 \\ 2s_1 + s_2 + 2s_3 \\ 2s_1 + s_2 + 2s_3 + 2s_4 \end{array} & \left[\begin{array}{cccccc} 1 & & & -1 & & \\ & 2 & & & -1 & \\ & & 2 & -2 & -1 & \\ & & & -2 & -1 & 2 \end{array} \right] \end{array}$$

(blank spaces in this and succeeding matrices are understood to be zero entries) from which we conclude that the mechanism is represented by

$$\mathbf{m} = \rho(2s_1 + s_2 + 2s_3 + 2s_4)$$

with ρ unrestricted. Since the number of cycles is zero, there is only one mechanism corresponding to the simple overall reaction.

In Horiuti's terminology (Horiuti and Nakamura, 1967) the "stoichiometric numbers" are $v_1 = v_3 = v_4 = 2$ and $v_2 = 1$, corresponding to the simple overall reaction.

Next we consider a slightly more complicated case in which a single cycle occurs, the hydrogen electrode reaction. This system has received considerable attention. Milner (1964) includes it in an interesting study of several electrode reactions in which the concept of direct paths is introduced. It also forms the basis of discussions by Horiuti (1973).

Example 3.5.2

The overall reaction of the system is given as

$$2H^+ + 2e^- = H_2$$

and three mechanistic steps are postulated, whose elementary reactions are given as

$$H^+ + H + e^- \rightleftharpoons H_2$$
$$H^+ + e^- \rightleftharpoons H$$
$$H + H \rightleftharpoons H_2$$

H is the only intermediate. Knowing this, we characterize the system as

$$R(s_1) = -H^+ - H - e^- + H_2$$
$$R(s_2) = -H^+ - e^- + H$$
$$R(s_3) = -2H + H_2$$

Here H^+ and e^- are always together which means that the system is electrically neutral and that $H^+ + e^-$ is a single chemical component. Therefore, let us omit the e^- from here on. Our method for finding all the direct mechanisms proceeds as follows.

The matrix corresponding to the preceding equation is

	H	H_2	H^+
s_1	−1	1	−1
s_2	1	0	−1
s_3	−2	1	0

3.5 Systems with a Simple Overall Reaction

Diagonalization gives

	H	H_2	H^+
s_2	1	0	-1
$s_1 + s_2$	0	1	-2
$s_1 - s_2 - s_3$	0	0	0

from which we conclude that space of mechanisms producing the overall reaction is represented by

$$\mathbf{m} = \rho(s_1 + s_2) + \phi(s_1 - s_2 - s_3)$$

with ρ and ϕ unrestricted. Collecting terms we obtain

$$\mathbf{m} = (\rho + \phi)s_1 + (\rho - \phi)s_2 - \phi s_3$$

To list the cycle-free subsystems, we determine the nonsingular 1×1 submatrices of the cycle-versus-step matrix

$$\begin{array}{cccc} & s_1 & s_2 & s_3 \\ \mathbf{m} & (1 & -1 & -1) \end{array}$$

There are three such submatrices (trivially), whose column headings are s_1, s_2, and s_3. The direct mechanisms are found by setting the coefficients of these steps equal to zero, one at a time, in $\mathbf{m}$:

$$\rho + \phi = 0, \quad \rho - \phi = 0, \quad -\phi = 0$$

Solving for ϕ in each case and substituting in $\mathbf{m}$, we get the following direct mechanisms:

$$m_1 = (2s_2 + s_3), \quad m_2 = (2s_1 - s_3), \quad m_3 = (s_1 + s_2)$$

This example will serve to illustrate the difference between our method and that employed in the pioneering studies by Horiuti and his co-workers (Horiuti and Ikusima, 1939; Horiuti and Nakamura, 1957, 1967; Horiuti, 1973). In Horiuti's terminology, for mechanism m_1 stoichiometric numbers are equal to $v_2 = 2$, $v_S = 1$; for m_2 we have $v_1 = 2$, $v_3 = -1$; and for m_3, $v_1 = 1$, $v_2 = 1$. However, Horiuti based his theory on finding the number of linearly independent mechanisms that can be combined algebraically to obtain all direct mechanisms. Thus in his solution of this problem, Horiuti found m_1 and m_3 and remarked that

$$m_2 = 2m_3 - m_1$$

He concluded that m_2 need not be considered as an additional basic mechanism. Miyahara (1969) and Horiuti (1973) noted that any two of the three mechanisms are algebraically independent. Even though any two can be combined algebraically to form the third, it must be noted they are each distinct chemically, in the sense that each of them involves a step that is not used at all in the third. Milner (1964) indicated this and presented a correct approach to finding the direct paths corresponding to a simple overall reaction. The following example involving two cycles is slightly more complicated.

Example 3.5.3

The ammonia synthesis reaction

$$N_2 + 3H_2 = 2NH_3$$

has been studied extensively from a mechanistic viewpoint (Horiuti, 1973; Temkin, 1973) and two quite different mechanisms have been proposed. Taking a more general viewpoint, we shall find that these two mechanisms belong to a family of six direct mechanisms for the ammonia synthesis reaction.

One of the mechanisms involves four mechanistic steps (proposed by Temkin) and the other involves five steps (proposed by Horiuti). Altogether there are nine distinct steps, which have been regarded as possibilities. Let us regard them as the complete list of steps. Then let us verify that the overall reaction space of the total system is the same as it was for the two original mechanisms, and let us determine the list of direct mechanisms that produce the reaction.

In the following list of elementary reactions for the ammonia synthesis reactions (1), (2), (3), and (7) were proposed by Temkin and (4), (5), (6), (8), and (9) by Horiuti:

(1) $\quad N_2 + l \rightleftharpoons N_2 l$

(2) $\quad N_2 l + H_2 \rightleftharpoons N_2 H_2 l$

(3) $\quad N_2 H_2 l + l \rightleftharpoons 2NHl$

(4) $\quad N_2 + 2l \rightleftharpoons 2Nl$

(5) $\quad Nl + Hl \rightleftharpoons NHl + l$

(6) $\quad NHl + Hl \rightleftharpoons NH_2 l + l$

(7) $\quad NHl + H_2 \rightleftharpoons NH_3 + l$

(8) $\quad H_2 + 2l \rightleftharpoons 2Hl$

(9) $\quad NH_2 l + Hl \rightleftharpoons NH_3 + 2l$

3.5 Systems with a Simple Overall Reaction

The symbol l in these elementary reactions refers to active surface sites on the catalyst. Every species with an l in it is an intermediate, and the rest are terminal species. For the purposes of our analysis we could omit l wherever it appears alone as a reactant. It is often convenient to use it to identify the intermediates. By including it as an intermediate in the present example, we get a case of a system where the maximum number M of steps in any direct mechanism is actually less than $I + R$ (the number of intermediates plus the dimension of the overall reaction space). The general rule is $M \leq I + R$.

The step-versus-species matrix for the present chemical system is

	N_2l	N_2H_2l	NHl	Nl	Hl	NH_2l	l	N_2	H_2	NH_3
s_1	1						-1	-1		
s_2	-1	1							-1	
s_3		-1	2				-1			
s_4				2			-2	-1		
s_5			1	-1	-1		1			
s_6			-1		-1	1	1			
s_7			-1				1	-1	1	
s_8					2		-2	-1		
s_9					-1	-1	2			1

Next, the matrix is diagonalized:

	N_2l	N_2H_2l	NHl	Nl	Hl	NH_2l	l	N_2	H_2	NH_3
s_1	1						-1	-1		
$s_2 + s_1$		1					-1	-1	-1	
$s_3 + s_2 + s_1$			2				-2	-1	-1	
s_4				2			-2	-1		
$2s_5 + s_4 - s_3 - s_2 - s_1$					-2		2		1	
$2s_6 - 2s_5 - s_4 + 2s_3 + 2s_2 + 2s_1$						2	-2	-1		-2
$2s_7 + s_3 + s_2 + s_1$							-1	-3	2	
$s_8 + 2s_5 + s_4 - s_3 - s_2 - s_1$										
$s_9 + s_8 - s_7 + s_6$										

The seventh row shows that there is a simple overall reaction, and the last two rows show that the cycle has a dimension of 2. The general mechanism is

$$\rho(2s_7 + s_3 + s_2 + s_1) + \phi(s_8 + 2s_5 + s_4 - s_3 - s_2 - s_1)$$
$$+ \psi(s_9 + s_8 - s_7 + s_6)$$

with ρ, ϕ, and ψ unrestricted. Collecting terms, we obtain

$$(\rho - \phi)(s_1 + s_2 + s_3) + \phi(s_4 + 2s_5) + \psi(s_6 + s_9) + (2\rho - \psi)s_7 + (\phi + \psi)s_8$$

Notice that the combinations of steps $(s_1 + s_2 + s_3)$, $(s_4 + 2s_5)$, and $(s_6 + s_9)$ can now be regarded as single steps. Accordingly there are five steps, two of which must be taken away to form a cycle-free subsystem (since the cycle space has dimension 2).

The total number of possibilities for removing two steps from the five shown in the general mechanism is given by the combination

$$\binom{S}{C} = \frac{5!}{2!(5-2)!} = 10$$

This is not a large number of possible trials, and we could proceed by trial and error to set steps in the general mechanism equal to zero.

Taking the first term in the general mechanism $(\rho - \phi)(s_1 + s_2 + s_3)$ equal to zero, we set the factor $(\rho - \phi) = 0$ and obtain $\phi = \rho$. There are four remaining terms in the general mechanism. If we take $\phi(s_4 + 2s_5)$ equal to zero, it is necessary to take $\phi = 0$. Thus we obtain $\phi = \rho = 0$ and ψ is undetermined, so no mechanism results from this choice of a second step. Taking the next term $\psi(s_6 + s_9)$ equal to zero, we obtain $\psi = 0$. This choice constitutes a viable mechanism that is obtained by setting $\phi = \rho$ and $\psi = 0$ into the remaining three terms of the general mechanism that were not removed, as follows:

$$\rho(s_4 + 2s_5) + 2\rho s_7 + \rho s_8$$

All the eight remaining possible choices for elimination of two steps lead to values of ϕ and ψ except the one in which the two steps $\psi(s_6 + s_9) = 0$ and $(2\rho - \psi)s_7 = 0$ are chosen. However, three of the combinations lead to $\phi = 0$ and $\psi = 0$ so that only six of the ten possible choices lead to distinguishable mechanisms.

A more efficient procedure, which is also applicable in more complicated cases, is to construct the cycle matrix that contains the ten 2×2 submatrices:

	$s_1 + s_2 + s_3$	$s_4 + 2s_5$	$s_6 + s_9$	s_7	s_8
$s_8 + 2s_5 + s_4 - s_3 - s_2 - s_1$	-1	1			1
$s_9 + s_8 - s_7 + s_6$			1	-1	1

There are two singular 2×2 submatrices, corresponding to columns 1, 2 and columns 3, 4. The remaining eight 2×2 submatrices are nonsingular and correspond to the following pairs of steps. Beside each pair are the values that must be taken by ϕ and ψ in the general mechanism, if that pair of steps is

3.5 Systems with a Simple Overall Reaction

absent (zero degree of advancement):

$$s_1 + s_2 + s_3, s_6 + s_9; \quad \phi = \rho, \quad \psi = 0$$
$$s_1 + s_2 + s_3, s_7; \quad \phi = \rho, \quad \psi = 2\rho$$
$$s_1 + s_2 + s_3, s_8; \quad \phi = \rho, \quad \psi = -\rho$$
$$s_4 + 2s_5, s_6 + s_9; \quad \phi = 0, \quad \psi = 0$$
$$s_4 + 2s_5, s_7; \quad \phi = 0, \quad \psi = 2\rho$$
$$s_4 + 2s_5, s_8; \quad \phi = 0, \quad \psi = 0$$
$$s_6 + s_9, s_8; \quad \phi = 0, \quad \psi = 0$$
$$s_7, s_8; \quad \phi = -2\rho, \quad \psi = 2\rho$$

The direct mechanisms are found by substituting the values of ϕ and ψ into the general mechanism. The sixth and seventh pairs of values may be omitted, since they are repetitious of the fourth pair. Accordingly, there are six direct mechanisms that produce the overall reaction and they are

$$m_1 = \rho(s_4 + 2s_5) + 2\rho s_7 + \rho s_8$$
$$m_2 = \rho(s_4 + 2s_5) + 2\rho(s_6 + s_9) + 3\rho s_8 \quad \text{(Horiuti)}$$
$$m_3 = \rho(s_4 + 2s_5) - \rho(s_6 + s_9) + 3\rho s_7$$
$$m_4 = \rho(s_1 + s_2 + s_3) + 2\rho s_7 \quad \text{(Temkin)}$$
$$m_5 = \rho(s_1 + s_2 + s_3) + 2\rho(s_6 + s_9) + 2\rho s_8$$
$$m_6 = 3\rho(s_1 + s_2 + s_3) - 2\rho(s_4 + s_5) + 2\rho(s_6 + s_9)$$

For comparison these mechanisms can be tabulated as follows with the common factor of ρ ignored;

	$(s_1 + s_2 + s_3)$	$(s_4 + 2s_5)$	$(s_6 + s_9)$	s_7	s_8	
m_1	0	1	0	2	1	
m_2	0	1	2	0	3	(Horiuti)
m_3	0	1	-1	3	0	
m_4	1	0	0	2	0	(Temkin)
m_5	1	0	2	0	2	
m_6	3	-2	2	0	0	

The symbols m_2 and m_4 represent the mechanisms proposed by Horiuti and Temkin, respectively; m_3 and m_6 might be omitted on the grounds that some of their steps proceed in the reverse of an accepted direction, but still m_1 and m_5 then remain for consideration.

In the examples thus far considered in this section, we started with the assumption that a mechanism or mechanisms were to be found corresponding to a simple overall reaction. If sufficient mechanistic steps had not been postulated, it would not have been possible to find mechanisms corresponding to the assumed overall reaction. When more than one simple overall reaction is possible, an atom by species matrix can be employed to determine the maximum dimension of the overall reaction space (i.e., the number of independent overall reactions using customary terminology). Thus in Example 3.1.2 it was shown that in the case of ethylene oxide the maximum dimension of the overall reaction space is two. However, if we were to assume a small enough number of intermediate steps, it would be possible to obtain only a simple overall reaction of dimension unity.

The method of calculation that we have employed will give the actual dimension of the overall reaction space consistent with the choice of mechanistic steps that are assumed. The following example illustrates this situation.

Example 3.5.4

Suppose the following steps are postulated for the oxidation of ethylene to produce ethylene oxide (Lyubarski, 1956). It is desired to determine whether any mechanisms exist that correspond to these steps and also to find the dimension of the overall reaction, if one exists.

(1) $\quad O_2 + l \rightleftharpoons O_2 l$

(2) $\quad O_2 l + C_2 H_4 \rightleftharpoons C_2 H_4 O + O l$

(3) $\quad O l + C_2 H_2 \rightleftharpoons CH_2 CHOH l$

(4) $\quad 2 CH_2 CHOH l + 5 O_2 \rightleftharpoons 4 CO_2 + 4 H_2 O + 2 l$

The matrix of coefficients is

	$O_2 l$	$O l$	$CH_2 CHOH l$	O_2	$C_2 H_4$	$C_2 H_4 O$	CO_2	$H_2 O$
s_1	1			-1				
s_2	-1	1			-1	1		
s_3		-1	1		-1			
s_4			-2	-5			4	4

The diagonalized matrix is

	$O_2 l$	$O l$	$CH_2 CHOH l$	O_2	$C_2 H_4$	$C_2 H_4 O$	CO_2	$H_2 O$
s_1	1			-1				
$s_1 + s_2$		1		-1	-1	1		
$s_1 + s_2 + s_3$			1	-1	-2	1		
$2s_1 + 2s_2 + 2s_3 + s_4$				-7	-4	2	4	4

from which we conclude that the mechanism is represented by

$$\mathbf{m} = \rho(2s_1 + 2s_2 + 2s_3 + s_4)$$

The number of cycles is zero and thus there is only one overall reaction,

$$7O_2 + 4C_2H_4 = 2C_2H_4O + 4CO_2 + 4H_2O$$

This mechanism would correspond, in Horiuti's terminology, to taking stoichiometric numbers of the steps as

$$v_1 = v_2 = v_3 = 2; \quad v_4 = 1.$$

Such a mechanism would correspond to a constant selectivity for ethylene oxide formation of 50%. Such a conclusion does not agree with observed data for this system. It serves, however, to illustrate how fewer overall reactions can result, if more restricted choices of steps are assumed.

3.6 Systems with a Multiple Overall Reaction

The chemical systems considered in this section have an overall reaction space with a dimension greater than one. This is developed in the course of the procedure for enumerating mechanisms. If it were desired to determine whether this were the maximum dimension possible consistent with the selection of terminal species, it would be necessary to make separate atom-by-species matrix calculations, as discussed in Section 3.1.

In many industrial reactions involving organic substances a major product will be formed, but a side reaction will contribute to loss in selectivity of yield of the desired product. In such cases the mechanism will involve a multiple overall reaction. Typical of this is the oxidation of ethylene oxide, which is considered in Example 3.5.4, but with enough steps so that two independent overall reactions occur. This reaction system has been considered both by Miyahara (1969) and Temkin (1979). We will illustrate the procedure recommended here by using illustrative examples for the system.

The procedure used by Miyahara for selecting mechanistic steps follows that previously discussed wherein intermediates are chosen so that one single step involves the production of an intermediate and another single step involves its consumption. For such a selection, the sum of the number of independent intermediates and the dimension of the overall reaction will just equal the number of mechanistic steps and there will be no cycles. In our procedure, this corresponds to a single direct mechanism for the system. Two sets of stoichiometric numbers could be used to designate the number of times each step is taken for each of the simple overall reactions chosen, but they are not unique as in the case of a one-dimensional overall reaction. Nevertheless, if the rates of the two chosen independent reactions are specified, that is sufficient information to fix the system and the rates of all the steps in the single direct mechanism.

Example 3.6.1

Miyahara's proposed system for ethylene oxide formation is as follows:

(1) $O_2 + 2l \rightleftharpoons 2Ol$

(2) $C_2H_4 + Ol \rightleftharpoons C_2H_4Ol$

(3) $CH_2Ol + 2l \rightleftharpoons COl + 2Hl$

(4) $COl + Ol \rightleftharpoons CO_2 + 2l$

(5) $2Hl + Ol \rightleftharpoons H_2O + 3l$

(6) $C_2H_4Ol \rightleftharpoons C_2H_4O + l$

(7) $C_2H_4 + 2Ol \rightleftharpoons 2CH_2Ol$

The intermediates in this system are the species involving l and the others are terminal species. We construct the step-versus-species matrix:

	Ol	C_2H_4Ol	CH_2Ol	COl	Hl	l	O_2	C_2H_4	C_2H_4O	CO_2	H_2O
s_1	2					-2	-1				
s_2	-1	1						-1			
s_3			-1	1	2	-2					
s_4	-1			-1		2				1	
s_5	-1				-2	3					1
s_6		-1				1			1		
s_7	-2		2					-1			

Next, we diagonalize the matrix:

	Ol	C_2H_4Ol	CH_2Ol	COl	Hl	l	O_2	C_2H_4	C_2H_4O	CO_2	H_2O
s_1	2					-2	-1				
$2s_2 + s_1$		2				-2	-1	-2			
s_3			-1	1	2	-2					
$2s_4 + s_1$				-2		2	-1			2	
$2s_5 + s_1$					-4	4	-1				2
$2s_6 + 2s_2 + s_1$							-1	-2	2		
$s_7 + 2(s_5 + s_4 + s_3) + 3s_1$							-3	-1		2	2

Since there are no cycles in this system, the general mechanism is a direct mechanism and is

$$\rho(s_1 + 2s_2 + 2s_6) + \sigma(3s_1 + 2s_3 + 2s_4 + 2s_5 + s_7)$$

3.6 Systems with a Multiple Overall Reaction

This produces a multiple overall reaction:

$$\rho(-O_2 - 2CH_2H_4 + 2C_2H_4O) + \sigma(-3O_2 - C_2H_4 + 2CO_2 + 2H_2O)$$

In both of these expressions ρ and σ are unrestricted, and, since there is more than one unrestricted coefficient, the overall reaction is described as *multiple*. The form in which this multiple overall reaction is written suggests that it is made up of two specific reactions, operating independently, but this separation is not inherent in the chemical system. The same multiple overall reaction is represented, for instance, by

$$\rho'(-O_2 - 2C_2H_4 + 2C_2H_4O) + \sigma'(-4CO_2 - 2H_2O + 2C_2H_4O + 5O_2)$$

in which ρ' and σ' are unrestricted. The method of representation is arbitrary. By collecting terms, we can write the general mechanism and its multiple overall reaction as

$$(\rho + 3\sigma)s_1 + 2\rho s_2 + 2\sigma s_3 + 2\sigma s_4 + 2\sigma s_5 + 2\rho s_6 + \sigma s_7$$

and

$$(\rho + 3\sigma)O_2 + (2\rho + \sigma)C_2H_4 \rightleftharpoons 2\rho C_2H_4O + 2\sigma CO_2 + 2\sigma H_2O$$

This conclusion is essentially the same as that of Miyahara.

In accordance with Horiuti's rule that the number of linearly independent routes is equal to the number of steps minus the number of linearly independent intermediates (7–5), there are two independent routes corresponding to two overall reactions.

Example 3.6.2

A more complicated case is treated here following Temkin's proposed system for ethylene oxide formation. This system has the same overall reaction space as Example 3.6.1, but has some differences of intermediates. Notice, in particular, that CH_3CHO is an intermediate. The other intermediates are those species bound to a catalyst, indicated by the letter l in the formula. The elementary reactions are

(1) $\quad O_2 + l \rightleftharpoons O_2 l$

(2) $\quad 2Ol \rightleftharpoons l + O_2$

(3) $\quad O_2 l + C_2H_4 \rightleftharpoons Ol + CH_3CHO$

(4) $\quad C_2H_4O + l \rightleftharpoons C_2H_4Ol$

(5) $\quad O_2 l + C_2H_4 \rightleftharpoons C_2H_4O + Ol$

(6) $\quad 5O_2 l + CH_3CHO \rightleftharpoons 5Ol + 2CO_2 + 2H_2O$

(7) $\quad C_2H_4Ol \rightleftharpoons Ol + C_2H_4$

The matrix of coefficients is

	O_2l	Ol	CH_3CHO	C_2H_4Ol	l	O_2	C_2H_4	C_2H_4O	CO_2	H_2O
s_1	1				-1	-1				
s_2	1	-2				1				
s_3	-1	1	1				-1			
s_4				1	-1			-1		
s_5	-1	1				-1	1			
s_6	-5	5	-1						2	2
s_7		1			-1		1			

The diagonalized matrix is

	O_2l	Ol	CH_3CHO	C_2H_4Ol	l	O_2	C_2H_4	C_2H_4O	CO_2	H_2
s_1	1				-1	-1				
$s_2 - s_1$		-2				2	1			
$2s_3 + s_1 + s_2$			2				-1	-2		
s_4				1	-1			-1		
$s_1 + s_2 + 2s_5$						-1	-2		2	
$3s_1 + 3s_2 + s_3 + s_6$						-3	-1		2	2
$s_2 + s_4 + s_5 + s_7$										

The general mechanism is

$$\rho(s_1 + s_2 + 2s_5) + \sigma(3s_1 + 3s_2 + s_3 + s_6) + \phi(s_2 + s_4 + s_5 + s_7)$$

and the overall reaction is

$$\rho(-O_2 - 2C_2H_4 + 2C_2H_4O) + \sigma(-3O_2 - C_2H_4 + 2CO_2 + 2H_2O)$$

The general mechanism can be written as

$$(\rho + 3\sigma)s_1 + (\rho + 3\sigma + \phi)s_2 + \sigma(s_3 + s_6) + (2\rho + \phi)s_5 + \phi(s_4 + s_7)$$

Since there is only one cycle, $s_2 + s_4 + s_5 + s_7$, the values of ϕ, which determine the direct mechanism, are found by setting the coefficients of s_2, s_5, and $(s_4 + s_7)$ equal to zero in the general mechanism, which gives

$$\phi = -\rho - 3, \qquad \phi = -2\rho, \qquad \phi = 0$$

Placing these values in the general mechanism, we get the direct mechanisms:

$$m_1 = (\rho + 3\sigma)s_1 + \sigma(s_3 + s_6) + (\rho - 3\sigma)s_5 - (\rho + 3\sigma)(s_4 + s_7)$$

$$m_2 = (\rho + 3\sigma)s_1 + (-\rho + 3\sigma)s_2 + \sigma(s_3 + s_6) - 2\rho(s_4 + s_7)$$

$$m_3 = (\rho + 3\sigma)s_1 + (\rho + 3\sigma)s_2 + \sigma(s_3 + s_6) + 2\rho s_5$$

3.6 Systems with a Multiple Overall Reaction 65

For comparison these can be tabulated as follows:

	s_1	s_2	$s_3 + s_6$	s_5	$s_4 + s_7$
m_1	$\rho+3\sigma$	0	σ	$\rho-3\sigma$	$-\rho-3\sigma$
m_2	$\rho+3\sigma$	$-\rho+3\sigma$	σ	0	-2ρ
m_3	$\rho+3\sigma$	$\rho+3\sigma$	σ	2ρ	0

Temkin (1979) discusses this example. He obtains the double overall reaction and the cycle. He concludes that the number of basic routes (7–4) is equal to three, while there are two single overall reactions. By eliminating the route corresponding to the cycle or empty route, he arrives at a mechanism involving two overall reactions that is not unique. The mechanism he obtains corresponds to m_1 in our example and, in fact, is exhibited directly in the table showing the diagonalized matrix. However, m_2 and m_3 are equally valid choices as possible mechanisms. To proceed further, other information would be needed. For example, if spectroscopic observation indicated the absence of C_2H_4Ol, mechanisms m_1 and m_2 could be ruled out; absence of isotopic oxygen scrambling would tend to eliminate mechanisms m_2 and m_3; complete absence of CH_3CHO would make all mechanisms listed unlikely.

Example 3.6.3 Dehydrogenation of Butane

Another example treated by Temkin (1971) can serve to illustrate our procedure. The following elementary reactions are proposed for the dehydrogenation of butane to produce butene and butadiene. For simplicity isomers are not distinguished.

(1) $\quad C_4H_{10} + l \rightleftharpoons C_4H_8l + H_2$

(2) $\quad C_4H_8l \rightleftharpoons l + C_4H_8$

(3) $\quad C_4H_8l \rightleftharpoons C_4H_6l + H_2$

(4) $\quad C_4H_6l \rightleftharpoons l + C_4H_6$

(5) $\quad C_4H_{10} + l + C_4H_6l \rightleftharpoons 2C_4H_8l$

The matrix of coefficients is

	C_4H_8l	C_4H_6l	l	C_4H_{10}	C_4H_8	C_4H_6	H_2
s_1	1		−1	−1			1
s_2	−1		1		1		
s_3	−1	1					1
s_4		−1	1			1	
s_5	2	−1	−1	−1			

The diagonalized matrix is

	C_4H_8l	C_4H_6l	l	C_4H_{10}	C_4H_8	C_4H_6	H_2
s_1	1		-1	-1			1
$s_3 + s_1$		1	-1	-1			2
$s_2 + s_1$				-1	1		1
$s_4 + s_3 + s_1$				-1		1	2
$s_5 + s_1 + s_3$							

The overall reaction is represented by

$$\rho(-C_4H_{10} + C_4H_8 + H_2) + \sigma(-C_4H_{10} + C_4H_6 + 2H_2)$$

and the general mechanism, which produces it, is

$$\rho(s_1 + s_2) + \sigma(s_1 + s_3 + s_4) + \phi(s_5 - s_1 + s_3)$$

or

$$(\rho + \sigma - \phi)s_1 + \rho s_2 + (\sigma + \phi)s_3 + \sigma s_4 + \phi s_5$$

Since there is only one cycle $s_5 - s_1 + s_3$, the values of ϕ, which determine the direct mechanisms, are found by setting the coefficients of s_1, s_3, and s_5 equal to zero in the general mechanism, which gives

$$\phi = \rho + \sigma, \qquad \phi = -\sigma, \qquad \phi = 0$$

Placing these values in the general mechanism, we get three direct mechanisms:

$$m_1 = \rho s_2 + (\rho + 2\sigma)s_3 + \sigma s_4 + (\rho + \sigma)s_5$$
$$m_2 = (\rho + 2\sigma)s_1 + \rho s_2 + \sigma s_4 - \sigma s_5$$
$$m_3 = (\rho + \sigma)s_1 + \rho s_2 + \sigma s_3 + \sigma s_4.$$

For comparison these can be tabulated

	s_1	s_2	s_3	s_4	s_5
m_1	0	ρ	$\rho + 2\sigma$	σ	$\rho + \sigma$
m_2	$\rho + 2\sigma$	ρ	0	σ	$-\rho$
m_3	$\rho + \sigma$	ρ	σ	σ	0

In this example Temkin's solution amounts to a choice of mechanism m_3 as a basis on which to develop kinetic equations. As is seen from the table, each

3.6 Systems with a Multiple Overall Reaction

mechanism omits one step. If m_3 is chosen, it is tacitly assumed that the exchange reaction s_5 does not occur. If that were not assumed, then m_1 and m_2 would be alternative mechanisms to consider.

A more complicated system involving three cycles is considered in the following example. Here it is almost essential to use the procedure described in Section 3.5 involving the use of cycle matrices because of the large number of cases to be considered.

Example 3.6.4 Methanation of Synthesis Gas

A 16-step system is selected that includes many features discussed in the literature. It is by no means exhaustive but does illustrate the main possibilities for reaction mechanisms. The idea of a CHO_{ads} intermediate has been advocated by Pichler and Schultz (1970), although others such as $CHOH_{ads}$ could have been equally well postulated to take into account the hypothesis that CO dissociation can be assisted by hydrogen. The water-gas shift reaction is considered to occur via the $CHOO_{ads}$ intermediate following studies of Oki et al. (1972). The hydrogenation of carbidic species by atomic rather than molecular hydrogen follows the findings of Happel et al. (1981). The following elementary reactions are assumed:

(1) $COl + l \rightleftharpoons Cl + Ol$

(2) $Cl + H_2 \rightleftharpoons CHl + l$

(3) $CHl + Hl \rightleftharpoons CH_2l + l$

(4) $CH_2l + Hl \rightleftharpoons CH_3l + l$

(5) $CH_3l + Hl \rightleftharpoons CH_4 + 2l$

(6) $OHl + Hl \rightleftharpoons H_2O + 2l$

(7) $CO_2 + l \rightleftharpoons CO_2l$

(8) $CO + l \rightleftharpoons COl$

(9) $H_2 + 2l \rightleftharpoons 2Hl$

(10) $CO_2l + Hl \rightleftharpoons HCOOl + l$

(11) $HCOOl + Hl \rightleftharpoons CHOl + OHl$

(12) $Ol + Hl \rightleftharpoons OHl + l$

(13) $COl + Ol \rightleftharpoons CO_2l + l$

(14) $HCOOl + l \rightleftharpoons OHl + COl$

(15) $CHOl + Hl \rightleftharpoons CHl + OHl$

(16) $COl + Hl \rightleftharpoons CHOl + l$

The step-by-species matrix is

	Ol	Cl	CHl	CH$_2l$	CH$_3l$	OHl	CO$_2l$	COl	Hl	CHOOl	CHOl	CH$_4$	CO$_2$	CO	H$_2$O	H$_2$
s_1	1	1														
s_2		−1	1													
s_3			−1	1												
s_4				−1	1											
s_5					−1							1				
s_6						−1			−1						1	
s_7							1		−1							
s_8								1	−1				−1			
s_9									2							−1
s_{10}							−1		−1	1						
s_{11}						1			−1	−1	1					
s_{12}	−1					1			−1							
s_{13}	−1						1	−1								
s_{14}								1			−1	1				
s_{15}						1			−1		−1					
s_{16}								−1	−1			1				

Diagonalization of this matrix results in the following:

	O*l*	C*l*	CH*l*	CH₂*l*	CH₃*l*	OH*l*	CO₂*l*	CO*l*	H*l*	CHOO*l*	CHO*l*	CH₄	CO₂	CO	H₂O	H₂
s_1	1	1														
s_2		−1	1													
s_3			−1	1												
s_4				−1	1											
s_5					−1							1				
s_6						−1									1	
s_7							1		−1				−1			
s_8								1	−1					−1		
s_9									2		2		−2	2		−1
$2s_7+s_9+2s_{10}$										2			−2	2		−1
$2s_6+2s_7+3s_9+2s_{10}+2s_{11}$											2		−2	2	2	−3
(let $t = s_1+s_2+s_3+s_4+s_5$)																
$t+s_6+s_8+3s_9+s_{12}$												1	−1	−1	1	−3
$t-s_7+2s_8+2s_9+s_{13}$												1	1	−2	2	−2
$-2s_1-2s_2+2s_{10}+2s_{11}-4s_{12}+2s_{13}+2s_{15}$																
$s_{10}-s_{12}+s_{13}+s_{14}$																
$-s_{10}-s_{11}+s_{12}-s_{13}+s_{16}$																

3 Enumeration of Reaction Mechanisms

The general mechanism is

$$\rho(t + s_6 + s_8 + 3s_9 + s_{12}) + \sigma(t - s_7 + 2s_8 + 2s_9 + s_{13})$$
$$+ \phi(s_{10} - s_{12} + s_{13} + s_{14})$$
$$+ \psi(s_1 + s_2 - s_{10} - s_{11} + 2s_{12} - s_{13} - s_{15})$$
$$+ \chi(-s_{10} - s_{11} + s_{12} - s_{13} + s_{16})$$

and the overall reaction is

$$\rho(-3H_2 - CO + H_2O + CH_4) + \sigma(-2H_2 - 2CO + CO_2 + CH_4)$$

The general mechanism can be written as

$$(\rho + \sigma + \psi)(s_1 + s_2) + (\rho + \sigma)(s_3 + s_4 + s_5) + \rho s_6 - \sigma s_7 + (\rho + 2\sigma)s_8$$
$$+ (3\rho + 2\sigma)s_9 + (\phi - \psi - \chi)s_{10} - (\psi + \chi)s_{11} + (\rho - \phi + 2\psi + \chi)s_{12}$$
$$+ (\sigma + \phi - \psi - \chi)s_{13} + \phi s_{14} - \psi s_{15} + \chi s_{16}$$

Notice that the combination of steps $(s_1 + s_2)$ and $(s_3 + s_4 + s_5)$ can now be regarded as single steps. Accordingly, there are 13 steps, three of which must be taken away to form a cycle-free subsystem (since the cycle space has a dimension of 3).

The total number of possibilities for taking away two steps from the 13 shown is given by the combination

$$\binom{S}{C} = \binom{13}{3} = \frac{13!}{3!(13-3)!} = 286$$

We will employ the cycle matrix to reduce the number of trials, omitting those columns where the steps do not involve cycles. We then have the following $S \times C$ matrix:

	$s_1 + s_2$	s_{10}	s_{11}	s_{12}	s_{13}	s_{14}	s_{15}	s_{16}
$s_{10} - s_{12} + s_{13} + s_{14}$	0	1	0	−1	1	1	0	0
$s_1 + s_2 - s_{10} - s_{11} + 2s_{12} - s_{13} - s_{15}$	1	−1	−1	+2	−1	0	−1	0
$-s_{10} - s_{11} + s_{12} - s_{13} + s_{14}$	0	−1	−1	1	−1	0	0	1

The $S \times C$ matrix thus has the following possible trials

$$\binom{S}{C} = \frac{8!}{3!(8-3)!} = 56$$

We consider which of these 56 combinations corresponds to a singular

3.6 Systems with a Multiple Overall Reaction

determinant. For example, $(s_1 + s_2), s_{10}, s_{11}$ corresponds to the determinant:

$$\begin{vmatrix} 0 & 1 & 0 \\ 1 & -1 & -1 \\ 0 & -1 & -1 \end{vmatrix} = 2$$

which is nonsingular. If we omit $(s_1 + s_2)$, (s_{10}), and (s_{11}) from the general mechanism corresponding to the omission of steps involving them:

$$\rho + \sigma + \psi = 0$$

$$\phi - \psi - \chi = 0$$

$$\psi + \chi = 0$$

Solving these equations for ϕ, ψ, and χ, we obtain

$$\phi = 0, \quad \psi = -\rho - \sigma, \quad \chi = \rho + \sigma$$

Of the 56 trials, only 34 are found to involve nonsingular matrices. Of these, 19 are found to be repetitious of other steps, so that 15 direct mechanisms remain. These mechanisms with steps omitted and the corresponding values of ϕ, ψ, and χ are as follows:

m_1	$s_1 + s_2, s_{10}, s_{11}$;	$\phi = 0$,	$\psi = -\rho - \sigma$,	$\chi = \rho + \sigma$
m_2	$s_1 + s_2, s_{10}, s_{16}$;	$\phi = -\rho - \sigma$,	$\psi = -\rho - \sigma$,	$\chi = 0$
m_3	$s_1 + s_2, s_{11}, s_{13}$;	$\phi = -\sigma$,	$\psi = -\rho - \sigma$,	$\chi = \rho + \sigma$
m_4	s_{10}, s_{11}, s_{12};	$\phi = 0$,	$\psi = -\rho$,	$\chi = \rho$
m_5	s_{10}, s_{11}, s_{15};	$\phi = 0$,	$\psi = 0$,	$\chi = 0$
m_6	s_{10}, s_{12}, s_{16};	$\phi = -\rho$,	$\psi = -\rho$,	$\chi = 0$
m_7	s_{11}, s_{12}, s_{15};	$\phi = \rho$,	$\psi = 0$,	$\chi = 0$
m_8	s_{11}, s_{13}, s_{15};	$\phi = -\rho$,	$\psi = 0$,	$\chi = 0$
m_9	s_{12}, s_{13}, s_{14};	$\phi = 0$,	$\psi = -\rho - \sigma$,	$\chi = \rho + 2\sigma$
m_{10}	s_{12}, s_{13}, s_{16};	$\phi = -\rho - 2\sigma$,	$\psi = -\rho - \sigma$,	$\chi = 0$
m_{11}	s_{13}, s_{14}, s_{15};	$\phi = 0$,	$\psi = 0$,	$\chi = \sigma$
m_{12}	s_{13}, s_{14}, s_{16};	$\phi = 0$,	$\psi = \sigma$,	$\chi = 0$
m_{13}	$s_1 + s_2, s_{14}, s_{16}$;	$\phi = 0$,	$\psi = -\rho - \sigma$,	$\chi = 0$
m_{14}	s_{12}, s_{14}, s_{15};	$\phi = 0$,	$\psi = 0$,	$\chi = -\rho$
m_{15}	s_{12}, s_{14}, s_{16};	$\phi = 0$,	$\psi = -\rho/2$,	$\chi = 0$

The direct mechanisms are found by substituting these values of $\phi, \psi,$ and χ into the general mechanism. For example, for m_1

$$(\rho + \sigma - \rho - \sigma)(s_1 + s_2) = 0$$
$$(0 + \rho + \sigma - \rho - \sigma)s_{10} = 0$$
$$(-\rho - \sigma + \rho + \sigma)s_{11} = 0$$
$$(\rho - 0 - 2\rho - 2\sigma + \rho + \sigma)s_{12} = -\sigma s_{12}$$
$$(\sigma + 0 + \rho + \sigma - \rho - \sigma)s_{13} = \sigma s_{13}$$
$$0(s_{14}) = 0$$
$$-(-\rho - \sigma)s_{15} = (\rho + \sigma)s_{15}$$
$$(\rho + \sigma)s_{16} = (\rho + \sigma)s_{16}$$

Coefficients for steps 3–9, of course, do not change for any of the mechanisms since they do not contain the unrestricted parameters $\phi, \psi,$ and χ. An alternative procedure for systematic elimination of repetitious trials is given by Happel and Sellers (1983).

For comparison, the steps corresponding to the complete set of mechanisms can be tabulated as shown on p. 73. The maximum number of steps $I + R = 13$ is obtained for mechanisms $m_6, m_{11}, m_{12}, m_{14},$ and m_{15}. No direct mechanism requires less than eight steps, as obtained in the case of mechanisms $m_1, m_3, m_5, m_9,$ and m_{10}. Of these, m_5 is probably the simplest since it requires no steps involving either CHOl or CHOOl, but it would be of interest to obtain data to discriminate between m_5 and the four other mechanisms that contain only eight steps.

In this chapter a method has been presented that furnishes a logical basis for modeling kinetic data to discriminate between possible mechanisms. A direct way of using this information involves the use of tracers since such modeling provides access to the step velocities involved. The next chapter shows how tracers may be applied in principle for this purpose. Of course, conventional methods of modeling using overall reaction kinetics can also benefit by the listing of possible direct mechanisms for modeling discussed in this chapter.

List of Symbols

A	Matrix of atomic coefficients
A	Number of species in a chemical system
A_n	A set of chemical elements with $n = 1, \ldots, N$
a_i	ith species in a chemical system

	s_1	s_2	s_3	s_4	s_5	s_6	s_7	s_8	s_9	s_{10}	s_{11}	s_{12}	s_{13}	s_{14}	s_{15}	s_{16}
m_1	0	0	$\rho+\sigma$	$\rho+\sigma$	$\rho+\sigma$	ρ	$-\sigma$	$\rho+2\sigma$	$3\rho+2\sigma$	0	0	$-\sigma$	σ	0	$\rho+\sigma$	$\rho+\sigma$
m_2	0	0	$\rho+\sigma$	$\rho+\sigma$	$\rho+\sigma$	ρ	$-\sigma$	$\rho+2\sigma$	$3\rho+2\sigma$	0	$\rho+\sigma$	$-\sigma$	0	$-\rho-\sigma$	$\rho+\sigma$	0
m_3	0	0	$\rho+\sigma$	$\rho+\sigma$	$\rho+\sigma$	ρ	$-\sigma$	$\rho+2\sigma$	$3\rho+2\sigma$	$-\sigma$	0	0	0	$-\sigma$	$\rho+\sigma$	$\rho+\sigma$
m_4	σ	σ	$\rho+\sigma$	$\rho+\sigma$	$\rho+\sigma$	ρ	$-\sigma$	$\rho+2\sigma$	$3\rho+2\sigma$	0	0	0	σ	0	ρ	ρ
m_5	$\rho+\sigma$	$\rho+\sigma$	$\rho+\sigma$	$\rho+\sigma$	$\rho+\sigma$	ρ	$-\sigma$	$\rho+2\sigma$	$3\rho+2\sigma$	0	0	ρ	0	0	ρ	0
m_6	σ	σ	$\rho+\sigma$	$\rho+\sigma$	$\rho+\sigma$	ρ	$-\sigma$	$\rho+2\sigma$	$3\rho+2\sigma$	0	0	0	σ	$-\rho$	ρ	0
m_7	$\rho+\sigma$	$\rho+\sigma$	$\rho+\sigma$	$\rho+\sigma$	$\rho+\sigma$	ρ	$-\sigma$	$\rho+2\sigma$	$3\rho+2\sigma$	ρ	0	$\rho+\sigma$	$\rho+\sigma$	0	0	0
m_8	$\rho+\sigma$	$\rho+\sigma$	$\rho+\sigma$	$\rho+\sigma$	$\rho+\sigma$	ρ	$-\sigma$	$\rho+2\sigma$	$3\rho+2\sigma$	$-\sigma$	0	0	0	$-\rho$	0	0
m_9	0	0	$\rho+\sigma$	$\rho+\sigma$	$\rho+\sigma$	ρ	$-\sigma$	$\rho+2\sigma$	$3\rho+2\sigma$	$-\sigma$	$-\sigma$	0	0	$-\sigma$	0	0
m_{10}	$\rho+\sigma$	0	$\rho+\sigma$	$\rho+\sigma$	$\rho+\sigma$	ρ	$-\sigma$	$\rho+2\sigma$	$3\rho+2\sigma$	$-\sigma$	$\rho+\sigma$	$\rho+\sigma$	0	0	$\rho+\sigma$	$\rho+2\sigma$
m_{11}	$\rho+2\sigma$	$\rho+\sigma$	$\rho+\sigma$	$\rho+\sigma$	$\rho+\sigma$	ρ	$-\sigma$	$\rho+2\sigma$	$3\rho+2\sigma$	$-\sigma$	$-\sigma$	$\rho+\sigma$	0	$-\rho-2\sigma$	$\rho+\sigma$	0
m_{12}	0	$\rho+2\sigma$	$\rho+\sigma$	$\rho+\sigma$	$\rho+\sigma$	ρ	$-\sigma$	$\rho+2\sigma$	$3\rho+2\sigma$	$-\sigma$	$-\sigma$	$\rho+\sigma$	0	0	$-\sigma$	ρ
m_{13}	$\rho+\sigma$	0	$\rho+\sigma$	$\rho+\sigma$	$\rho+\sigma$	ρ	$-\sigma$	$\rho+2\sigma$	$3\rho+2\sigma$	$\rho+\sigma$	$\rho+\sigma$	$-\rho-2\sigma$	$\rho+2\sigma$	0	$\rho+\sigma$	0
m_{14}	$\rho+\sigma$	$\rho+\sigma$	$\rho+\sigma$	$\rho+\sigma$	$\rho+\sigma$	ρ	$-\sigma$	$\rho+2\sigma$	$3\rho+2\sigma$	ρ	ρ	0	$\rho+\sigma$	0	0	$-\rho$
m_{15}	$\rho/2+\sigma$	$\rho/2+\sigma$	$\rho+\sigma$	$\rho+\sigma$	$\rho+\sigma$	ρ	$-\sigma$	$\rho+2\sigma$	$3\rho+2\sigma$	$\rho/2$	$\rho/2$	0	$(\rho/2)+\sigma$	0	$\rho/2$	0

B_s	A set of chemical species with $B = 1, \ldots, S$
$C = S - H - R$	Dimension of cycle space
H	Rank of matrix of stoichiometric coefficients of intermediates only
I	Number of intermediates in a chemical system
l	Active site on a catalyst
m	Mechanism
m_i	ith element in a basis for mechanism space
N	Number of elements in a chemical system
P	Numer of relations between the columns of matrix A
$\mathbf{R}$	Linear transformation of mechanism space to reaction space
R	Dimension of overall reaction space
$\mathbf{R(m)}$	Reaction produced by mechanism m
$\mathbf{R}(z)$	Mechanistic vector of cycles
$r_i = R(s_i)$	ith elementary reaction in a chemical system
$\mathbf{r}$	Multiple overall reaction
S	Number of distinguishable species in a chemical system
s	Step in a mechanism
s_i	ith step (type of molecular interaction) in a chemical system
$T = A - I$	Number of terminal species in a chemical system
α_{ij}	Stoichiometric coefficient of species α_j in elementary reaction r_i
β_{ij}	Stoichiometric coefficient of species α_j in elementary $\mathbf{R}(m_i)$
γ_{ij}	Coefficient of advancement of step s_j in steady-state mechanism m_i
μ_i	Coefficient of m_i in a mechanism m
ρ	Rate of advancement of a mechanism
σ	Rate of advancement of a mechanism
σ_i	Coefficient of step s_i in a mechanism m
τ	Rate of advancement of a mechanism
ϕ	Rate of advancement of a cycle
χ	Rate of advancement of a cycle
ψ	Rate of advancement of a cycle

References

Aris, R. (1969). "Elementary Chemical Reactor Analysis." Prentice-Hall, Englewood Cliffs, New Jersey.

Aris, R., and Mah, R. H. S. (1963). *I&EC Fundam.* **2**, 90.

References

Birkhoff, G., and MacLane, S. (1967). "A Survey of Modern Algebra," 3rd Ed., p. 160. Macmillan, New York.
Björnborn, P. H. (1977). *AIChE J.* **23**, 285.
Happel, J., and Sellers. P. H. (1982). *Ind. Eng. Chem. Fundam.* **21**, 67.
Happel, J., and Sellers. P. H. (1983). *Adv. Catal.* **32**, 273.
Happel, J., Fthenakis, V., Suzuki, J., Yoshida, T., and Ozawa, S. (1981). *Proc. Int. Congr. Catal. 7th, Tokyo.*
Horiuti, J. (1973). *Ann. N.Y. Acad. Sci.* **213**, 5.
Horiuti, J., and Ikusima, M. (1939). *Proc. Imp. Acad. (Tokyo)* **15**, 39.
Horiuti, J., and Nakamura, T. (1957). *Z. Phys. Chem., N.F.* **4**, 358.
Horiuti, J., and Nakamura, T. (1967). *Adv. Catal.* **17**, 1.
Hougen, O. A., and Watson, K. M. (1947). "Chemical Process Principles—Kinetics and Catalysis," Part 3, Wiley, New York.
Lyubarski, G. D. (1956). *Dokl. Nauk.* USSR **110**, 112.
Milner, P. C. (1964). *J. Electrochem. Soc.* **111**, 228.
Miyahara, K. (1969). *J. Res. Inst. Catal. Hokkaido Univ.* **17**, 219.
Oki, S., Happel, J., Hnatow, M. A., and Kaneko, Y. (1972). *Proc. Int. Congr. Catal., 5th.*
Pichler, H., and Schultz, H. (1970). *Chem. Ing. Tech.* **42**, 1162.
Sellers, P. H. (1983). *In* "Proceedings of the Symposium on Chemical Applications of Topology and Graph Theory," Vol. 28, pp. 420–429. (R. B. King, ed.), Elsevier, Amsterdam.
Temkin, M. I. (1971). *Int. Chem. Eng.* **11**, 709.
Temkin, M. I. (1973). *Ann. N.Y. Acad. Sci.* **213**, 79.
Temkin, M. I. (1979). *Adv. Catal.* **28**, 173.
Wei, J., and Prater, C. D. (1962). *Adv. Catal.* **13**, 206.

Chapter 4

Superposition Modeling of Mechanisms

4.1 Separation of Mechanism and Rate Equations

In applying tracers, each of the intermediates and terminal species is considered to reside in a separate compartment for purposes of mathematical analysis. In this chapter, we will discuss how the appropriate material balance equations are constructed and used to obtain the desired model parameters characterizing assumed mechanisms. In this procedure it is not necessary to develop expressions for the kinetics of the overall reaction rates.

In most studies devoted to determination of reaction mechanisms, kinetic data are employed, which involve measurement of reaction rates as functions of temperature, pressure, and concentrations of ambient species. In order to interpret such rate data a plausible mechanism must first be chosen involving a network of simple reactions, which when combined satisfy the required stoichiometry. Next a set of appropriate rate equations for these steps must be assumed, in which rates are expressed as functions of surface concentrations, which in turn must be related to the vapor-phase concentrations of reacting species. Unknown parameters consist of the surface concentrations of various reacting species and the reaction velocity constants, which relate reaction step velocities to surface or bulk concentrations. These parameters are determined by means of computer programs that compare predicted with observed overall rates. Unoccupied sites must be determined by separate experiments. In order to develop such equations various reaction-rate expressions have been proposed as discussed in Section 1.2.

The method employed here was developed in the course of our studies using tracers. It does not require the use of such kinetic equations and consequently dispenses with a number of assumptions.

In all cases, it is assumed that the system to be modeled is at steady state and, therefore, a brief review of the implications of this hypothesis is cardinal in importance.

4.2 The Steady-State Hypothesis

In the analysis of systems of heterogeneous catalytic reactions we note that all surface intermediates are constant at steady state. As Temkin (1979) has pointed out, catalytic reactions in industrial reactors are, as a rule, operated under steady-state conditions. Laboratory studies of reactions at steady-state conditions have the advantage of much simpler mathematical analysis of results as compared to nonsteady-state processes.

A reaction system is considered to be at steady state if the concentrations of all species in each element of the reaction space (i.e., volume in the case of a homogeneous reaction or the vapor space occupied by reactants and products as well as the catalytic surface in the case of heterogeneous reactions) do not change with time. Such a situation is rigorously possible only in an open system such as a tubular reactor or a circulation flow system. The species participating in elementary reactions that comprise a complex reaction system may be grouped into two categories: intermediates and reaction participants. At steady-state reaction regime the concentration of an intermediate does not change in time. If the intermediate consists of a chemisorbed species the rate of formation of this species in elementary reactions is equal to the rate of its consumption by other elementary reactions. The concentration of a reactant in the vapor space of the reactor does not change in time because along with formation and consumption in elementary reactions this species enters or leaves a volume element at a constant rate. If a species that is not present in the feed to the reactor is produced in the course of reaction, such a species must be considered as a product unless its rate of consumption is so rapid that its concentration in the system is negligible, as discussed in Section 2.2.

There are certain situations for which this hypothesis will be invalid. If a catalyst is rapidly degraded by formation of carbon deposits, as is the case with catalytic cracking, or if the feed to a catalytic reactor changes rapidly, as in catalyst systems designed to convert automotive exhaust, the catalyst will not be at steady state during normal use. Consideration of such systems lies outside the scope of the present treatment.

It should be noted, however, that in many cases which are conventionally treated as if the catalyst composition does not change, such is not actually the case. Catalyst composition may depend on such factors as water content of the reacting gases or on their oxidizing capacity. Kinetic experiments, which must be conducted by changing partial pressures of reactants, may also inadvertently change chemical composition of the catalyst. This problem is avoided by use of tracers.

An aspect of the steady-state hypothesis that has received a great deal of attention in the literature is its application to closed, constant volume reactors. It does not apply rigorously in these situations, but when it can be employed considerable simplification in mathematical interpretation of data is possible. In the case of homogeneous reactions involving several steps it is assumed that after the reaction has been in progress for a short time the rate of formation of certain intermediates equals their rate of disappearance. In that case, by setting the derivatives of concentrations with respect to time of such species equal to zero, it is possible to obtain algebraic equations that can be used to eliminate these concentrations. The mathematics of this hypothesis has been discussed by Aris (1969) as an instance of singular perturbation theory with special attention to its application in enzyme kinetics. Rice (1960) approached the problem of steady-state validation by a criterion that used a set of inequalities involving kinetic quantities and applied it to examples of branching-chain reactions as well as to the classic case of hydrogen bromide formation from hydrogen and bromine. The latter reaction led Bodenstein to introduce this hypothesis into chemical kinetics over 70 years ago.

Generally the steady-state hypothesis applies to reaction systems being studied in heterogeneous catalysis without the approximations necessary when it is applied to homogeneous reaction systems. There are cases where slow deposition of carbon or change in catalyst composition over a period of time may require special consideration to ensure that experimentation is conducted in a short enough time frame so as to avoid such complications.

4.3 Methods of Tracer Experimentation

We will assume that a number of mechanisms are available for testing, as a result of following some type of systematic procedure such as that developed in the last chapter. As a working hypothesis it will also be assumed that for the range of operating conditions under consideration one of these mechanisms predominates. If more than one mechanism exists, of course, the problem will be more complicated. Such situations can occur in practical contexts such as hydrocarbon cracking, where homogeneous and catalytic reactions can occur simultaneously. In such cases, it will often be possible to conduct experiments so that these effects can be separated.

4.3 Methods of Tracer Experimentation

When we conduct chemical reactions in a system at constant temperature, pressure, and concentrations of observed terminal components, the basic kinetic information obtained constitutes the rates of production or consumption of these terminal species after expiration of sufficient time to secure steady-state conditions. These rates are not all independent. As discussed in Chapter 3, it is often convenient to express them in terms of rates of advancement of chemical equations relating terminal species. If it is desired to do this, the choice of such chemical equations must also be restricted to those that are thermodynamically allowed at the observed partial pressures of reacting terminal species.

In addition to these overall kinetic data, we now wish to conduct pertinent experiments in which the species involved are marked by isotopes. The quantitative observation of surface concentrations of intermediates may be difficult, although their identification is important in the initial enumeration of possible mechanisms. Generally the only observations employed in modeling are the rates of change of isotopic markings in terminal species. If transients of isotopic markings of intermediates can be obtained by suitable instrumentation, it would be possible to model them as well.

It is possible to amass a considerable collection of data that must be in agreement with any mechanism to be selected from the possibilities. Thus, for example, let us consider the case of ethylene oxide synthesis where the following five terminal species are involved: O_2, C_2H_4, C_2H_4O, CO_2, and H_2O. If we choose ^{13}C as a tracer element, it would be possible, while the reaction is taking place at steady state, to mark either C_2H_4, C_2H_4O, or CO_2 and observe the rate of change of ^{13}C in the three species. Of course, all three could be marked simultaneously at different concentrations, if desired. This is a system with a multiple overall reaction, as discussed in Example 3.6.1. It would still be possible to obtain a single chemical reaction by combination of equations so as to omit one component such as O_2 or C_2H_4. Omission of a second component from the reacting mixture would result in a system in which a chemical reaction would not occur. Thus, we could study the transfer of ^{13}C from C_2H_4 to C_2H_4O in the presence or absence of water. Such transfers constitute exchange reactions. While they do not serve to characterize the overall system occurring at steady state it is clear that, if such exchanges are slow in the absence of competing adsorbates, they will also be slow under reaction conditions.

Similar experiments could be conducted using ^{18}O or D as tracers. In the case of ^{18}O tracing transfer could occur among the four terminal species O_2, $C_2H_4O_2$, CO_2, and H_2O and exchange reactions could be conducted among smaller numbers of species. Similarly, in the case of D, it would be possible to trace transfers among C_2H_4, C_2H_4O, and H_2O.

In conducting isotope transfer experiments, we need to consider whether transfer of isotopes can occur by elementary steps in addition to those involved in the chemical reaction system under study. Such steps can involve either steady-state transfer of isotopes by exchange reactions among terminal species or transient transfer of isotope to catalyst components and reaction intermediates. Mechanisms can be modeled to include such steps. We consider such possibilities separately in connection with developing equations for modeling.

In the methods to be discussed further in this chapter, we have adopted a more restricted point of view than in the previous chapter. Instead of identifying a mechanism only in terms of the sets enumerated, we wish to determine quantitatively the intermediate concentrations and reaction velocities of the mechanistic steps. Often, on thermodynamic grounds, it is possible to establish whether a reaction is reversible. If it is reversible all steps must be reversible, and the extents of reaction defined in the previous chapter refer to the differences between forward and reverse velocities of the corresponding steps. However, if a reaction is irreversible, one or more of the mechanistic steps must also be irreversible. For any such step that is irreversible the velocity in the reverse direction is zero, and mechanisms including such a step can be eliminated, if the degree of advancement required for that step is negative.

In preliminary study it is useful to conduct experiments in which the product species are traced. In any case where a product atom is not transferred to reactants containing the same atomic species, a path exists in which an irreversible step occurs. This assumes, of course, that there is an equal probability of all atoms of a kind in a given molecule being marked.

The most convincing type of isotope experimentation is conducted under conditions in which marked species are substituted for unmarked species, leaving all other conditions the same, so that in the absence of isotopic kinetic effects essentially the same catalytic reaction occurs both in the presence and absence of tracer. However, it is often possible to obtain much useful information by the use of procedures where this is not the case, as discussed in Section 1.4. In addition, there are several techniques involving isotopes that are especially useful in conjunction with the superposition technique emphasized here.

These techniques depend on the fact that some reaction systems involve series of irreversible reaction steps in which an ambient reactant converts a chemisorbed species. In such a case, if the terminal reactant species is removed by rapid flushing with an inert gas, such as helium, it can be assumed that the series of chemisorbed intermediates will remain unchanged although any species that are reversibly adsorbed will be removed. If a traced species is then substituted for the initial terminal reactant, the intermediates can be desorbed as marked species. This procedure was utilized (Happel et al., 1980) to determine the intermediates involved in the methanation of synthesis gas. After hydrocarbon intermediates had been formed, they reacted with deuterium and the deuteromethanes collected were assumed to correspond to the concentrations of chemisorbed intermediates. A variation of this technique consists of conducting the initial reaction with marked ^{13}CO so that the intermediates are marked. Then, by means of helium flushing, ^{13}CO is removed from the gaseous phase. The reaction is resumed using a mixture of hydrogen and unmarked CO. That the hydrogenation of intermediates is unidirectional is shown by the fact that no ^{13}CO is recovered in the effluent, although it appears in the product methane.

An interesting somewhat similar technique was employed by Biloen et al. (1979) for the Fischer–Tropsch reaction in which both methane and higher hydrocarbons were produced. In their experiments sub-monolayers of ^{13}C were deposited on the catalyst directly from carbon monoxide. These submonolayers were then contacted with $^{12}CO/H_2$ and the methane and higher hydrocarbons were analyzed for their ^{13}C contents. The proportion of ^{12}CO relative to ^{13}CO corresponded to very rapid completion of the monolayer by ^{12}C, so it was concluded that the step involving dissociation of CO_{ads} on the catalyst surface was rapid. Of course, this assumes that the intermediate obtained by predeposition corresponds to that produced in the course of the synthesis reaction itself.

Other variations are possible in which extraneous traced molecules are incorporated in the mixture of reactants and the extent of tracer incorporation into products is observed. Thus Schulz and Achtsnit (1976) studied the Fischer–Tropsch synthesis using a technique in which ethylene marked with ^{14}C was introduced into the H_2/CO feed mixture, and concluded that ethylene insertion was involved in the mechanism of higher hydrocarbon formation. In a classic investigation Kummer et al. (1948) added labeled ketene to synthesis gas and found that the carbon atom of the CH_2 group in CH_2CO was apparently not inserted into the growing chain. More recent studies by Ponec and van Barneveld (1979) with higher ketene concentrations indicate that insertion of the CH_2 group into ketene does occur. In experiments of this type it probably would be desirable to enumerate the possible mechanisms involved by reaction with the more complex set of terminal species.

It should also be noted here that the use of tracers, of course, does not preclude obtaining information using untraced components.

In this connection, an interesting strategy involves the determination of reaction rates close to initial conditions. This procedure has been employed in a different context by Yang and Hougen (1950). These authors used total pressure dependence of initial reaction rate to discriminate between models in which a single rate-determining step is assumed to apply throughout the concentration ranges of interest for reactants and products. The initial rate is for the reaction that occurs in the absence of products. In using this procedure, as applied to overall kinetic data, it is tacitly assumed that a product is not so strongly chemisorbed that the rate-determining step in a mechanism changes. In the case of tracer experimentation, it is possible to relax this restriction since there is no requirement to assume a rate-determining step. Instead, a preferred mechanism is assumed to occur. Then if the mechanism applies throughout the desired range of conversion, it is apparent that under experimentation at initial conditions, no step involving the reaction of a terminal product will correspond to an acceptable preferred mechanism. Thus in Example 3.5.3, illustrating mechanisms for ammonia formation, we could rule out the mechanism m_3 (since it involves dissociative adsorption of product ammonia; step 7), if ammonia production were substantial under initial conditions when only N_2 and H_2 are present in the reacting gas stream.

In tracer experiments it is necessary to employ an apparatus in which concentrations of terminal species and their rates of change can be observed. In some cases it is possible also to observe concentration changes in adsorbed species by employing surface observation techniques such as infrared spectroscopy. Depending on the information desired and the kinetics involved, three different schemes may be employed: use of plug-flow, constant volume, or continuous-stirred-type reactors. Measurements made are used in conjunction with modeling equations. In the case of steady-state tracing direct solution for the parameters is possible, but it is impossible to obtain concentrations of intermediates. With transient tracing it is necessary to solve the relevant differential equations using assumed values of the parameters, which are compared with observed tracer outputs. In the following section we consider the development of such equations.

4.4 General Considerations in Modeling with Tracers

In modeling heterogeneous catalysis systems, we seek to determine concentrations of all intermediate species including those that cannot be measured directly because they are adsorbed on the catalyst and to determine the velocities of reaction steps of the elementary reactions relating intermediates to terminal products.

The most general type of relationship connecting the state variables (concentrations of tracers or traced components) with the independent variables of time and distance through a catalyst bed is

$$\mathbf{g}\left(\mathbf{z}, \frac{\partial \mathbf{z}}{\partial w}, \frac{\partial \mathbf{z}}{\partial t}, \boldsymbol{\theta}\right) = 0 \tag{4.1}$$

where **g** is a vector of functions, **z** a vector of the state variables, **θ** a vector of the parameters, w the distance through a bed, and t the time.

We measure the values of model-state variables for reactions occurring in appropriate apparatus (stirred tank reactor, recirculating reactor, constant volume stirred tank, plug-flow reactor). If values can be obtained for the parameters, which when employed in the model equation enable the data to be predicted, we consider a given representation according to Eq. (4.1) to be a possible model. Generally it is not possible to express the parameters directly as functions of the independent variables and the observed concentration changes so that it is necessary to obtain solutions of the form:

$$\mathbf{z} = \mathbf{f}(w, t, \boldsymbol{\theta}) \tag{4.2}$$

Even when an explicit solution of Eq. (4.2) is not possible, numerical procedures can be employed that will enable the transients of **z** to be predicted, if appropriate values of **θ** are assumed. Statistical techniques can then be employed to obtain optimal values of the parameters that are consistent with the observed data. The text by Bard (1974) on nonlinear parameter estimation discusses details of numerical procedures to accomplish this. Examples are worked out in the field of chemical kinetics in which we seek to determine appropriate reaction velocity constants from the changes in concentrations of reacting species with time. Our problem is somewhat different, but the same basic methods are applicable. Bennett (1976) has also published an excellent survey of transient methods in heterogeneous catalysis, most of which are not, however, concerned with tracer applications.

Considerable simplification in modeling can be accomplished when the space variable is eliminated, with time as the only independent variable. Such modeling has been employed by theoretical biologists for a long time. Excellent surveys covering general mathematical theories and application of isotopic experiments to the kinetics of metabolites within complex systems have been published by Berman (1971, 1979). These studies often take the form of compartmental modeling in which a system is represented as a finite number of discrete interacting states. Particles within a single state are considered indistinguishable from each other and subject to the same forces and transition opportunities.

Biological systems are often highly nonlinear (reactions are not necessarily first order) and they are frequently not in a steady state. A great deal of simplification can be achieved in the experimental design and in the interpretation of the result through linear perturbations. As Berman has demonstrated, this results in the loss of information obtained in an experiment. For example, the description of changes in states due to several nonlinear processes may be of the form:

$$dx_1/dt = -k_1 x_1^2 + k_2 x_1 x_2 - k_3 \tag{4.3}$$

This equation contains zeroth- and second-order reaction terms. If we label the components of the system with a tracer, than each x_i can be represented as the sum of two parts x_i^- and x_i^*, where x_i^- is the stable or untraced material (Brownell et al., 1968) and x_i^* is the tagged material. Setting $x_i = x_i^- + x_i^*$ we obtain

$$\frac{d(x_1^- + x_1^*)}{dt} = -k_1(x_1^- + x_1^*)^2 + k_2(x_1^- + x_1^*)(x_2^- + x_2^*) - k_3 \quad (4.4)$$

When x_i^* is very small compared to x_i^-, it behaves as a tracer, and the equation for x_i^* may be written as

$$dx_1^*/dt = [-2k_1 x_1^- + k_2 x_2^- - (k_3/x_1^-)] x_1^* + (k_2 x_1^-) x_2^- \quad (4.5)$$

Equation (4.5) is a linear differential equation in the tracer variables x_i^* since they appear as first-order terms only. This demonstrates the well-known fact that a small perturbation of a nonlinear system behaves linearly. The coefficients in the equation are functions of the tracee (the nonlabeled material).

When the system is at a steady state the x_i^- are constant but the tracer equation reduces to a first-order differential equation with constant coefficients

$$dx_1^*/dt = -\lambda_{11} x_1^* + \lambda_{12} x_2^* \quad (4.6)$$

where

$$\lambda_{11} = [2k_1 x_1^- - k_2 x_2^- + k_3/x_1^-)] \quad \text{and} \quad \lambda_{12} = (k_2 x_1^-)$$

Under these conditions tracer data lead to the solution of a subsystem having new variables λ_{11} and λ_{12}. The relation between the λ's and k's and the x_i^-, as given in Eq. (4.5) and (4.6), are not apparent from the single-tracer experiment. They may be derived from a series of tracer experiments performed at various steady states.

Experiments of this type are convenient in biological studies in which radioactive tracers such as ^{14}C or ^3H are employed and can be detected at very small concentrations. In heterogeneous catalysis studies it is usually more convenient to employ stable isotopes. Detection of species containing such isotopes is easier if it is not necessary to employ them in very small concentrations. It is still possible to employ compartmental analysis in which the space variable is eliminated and to linearize the model Eq. (4.4.1). Instead of assuming that the fraction of tracer is small relative to the total concentration of reacting species, we assume that the total concentration tracer plus unmarked species remains constant, regardless of the level of marking.

Substitution of a traced species will cause no change in reaction rates, provided that isotopic kinetic effects are negligible. Advantage was first taken of this fact many years ago by Enemoto and Horiuti (1953) in steady-state tracing studies. We have employed it in both steady and transient studies, most recently in catalytic methanation (Happel et al., 1980, 1982; Otarod et al., 1983). In steady-state transfer, step velocities appear in linear algebraic equations involving tracer marking levels. Both step velocities and surface concentrations appear as parameters for transient transfer in sets of linear differential equations. Use of transient tracing has also been widely discussed by Le Cardinal et al. (1977), Happel (1978), and Le Cardinal (1978). Walter (1975) has also treated this problem in the general context of identifiability of the parameters.

Regardless of what method is employed to linearize the modeling equations, such a procedure effects considerable simplification in treatment. Linear systems have been widely studied for application to electric circuit and control theory, as well as in the biosciences.

In the section immediately following, the modeling equations, which are used with several types of reactors, will be derived and simple solutions will be developed to illustrate the form in which parameters appear. Computer programs can be employed to provide numerical estimates of parameters. The attempt to determine suitable values of parameters by such a procedure is a logical next step in modeling experimental data. Procedures for accomplishing this for steady-state and transient superposition will be discussed in following chapters.

However, at this point it is important to call attention to the fact that after obtaining estimates of parameters by such a procedure, it is necessary to further examine the reliability of the results obtained. We are thus led to considerations of structural identifiability and distinguishability. They will be only mentioned briefly at this point to emphasize their importance, if one is interested in ascribing more than empirical significance to parameters resulting from computer modeling. These subjects will be considered in detail in Chapter 6.

If the data that can be observed are insufficient to obtain the desired parameters corresponding to a given set of modeling equations, the model is said to be structurally unidentifiable. In that case, numerical solution may enable a set of parameters to be determined corresponding to one of a number of local solutions.

A closely related problem is that of discrimination. Two models may be proposed each of which could in principle predict the same transient response corresponding to given observed tracer outputs. The models would then be considered to be structurally indistinguishable.

4.5 Superposition Modeling with the Gradientless Reactor

Compartmental modeling of a reaction system generally leads to a set of equations in terms of fractional tracer marking z as the dependent variable

$$d\mathbf{z}/dt = \mathbf{A}\mathbf{z} + \mathbf{b} \tag{4.7}$$

The matrix **A** includes terms that allow for species in the gas phase as well as those adsorbed on the catalyst. It is assumed that each compartment corresponds to a single species, so the number of equations corresponds to the number of species. A material balance for tracer transient is made for each species or compartment.

To show how these equations are developed, let us first consider a simple three-compartment model (Happel and Kiang, 1977). Assume that a reaction is being conducted over a solid catalyst in a gradientless open reactor. Reactors of this type have been discussed in Section 1.3, and typical reactors are illustrated in Figs. 1.1–1.4. Such reactors operate at constant pressure and temperature with continuous input of feed and withdrawal of product. To illustrate the procedure, assume that the following reaction is conducted over a solid catalyst.

$$A \underset{v_{-1}}{\overset{v_{+1}}{\rightleftharpoons}} Al \underset{v_{-2}}{\overset{v_{+2}}{\rightleftharpoons}} B$$

where A and B are reactant and product that can be introduced or withdrawn from the system, Al is the chemisorbed intermediate, and $v_{\pm i}$ are mechanistic step velocities (moles/sec-g_{cat}). The system is shown schematically in Fig. 4.1.

It is assumed that either feed or product can be marked by an isotope corresponding to a single atom of an element present in the molecules involved. The reaction is conducted at steady state prior to the transient tracing experiment, and at time $t = 0$ a mixture containing marked species is substituted for the previously unmarked feed. The total molecules of feed and product remain unchanged, so that the overall reaction still occurs at the same steady state, assuming that no isotopic kinetic effects occur. The tracer redistribution that occurs is superimposed on this unchanged steady-state reaction. In terms of compartmental modeling there are three compartments corresponding to the three species A, Al, and B. A set of three differential

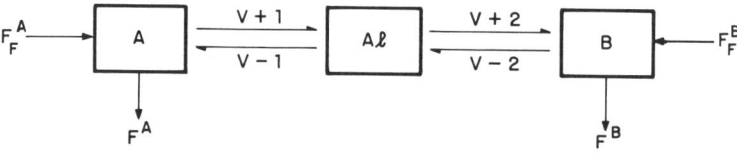

FIG. 4.1 Compartmental model-definition sketch.

4.5 Superposition Modeling with the Gradientless Reactor

equations may be written corresponding to material balances for tracer entering and leaving each compartment. Thus, for species A the input corresponds to $F^A z^A/W$, feed to the reactor per unit weight of catalyst, plus $v_{-1} z^{Al}$, tracer returning via the reverse of reaction step 1 from the catalyst surface. The corresponding output consists of $F^A z^A/W$, product leaving reactor plus $v_{+1} z^A$, tracer leaving the compartment A by reaction step 1.

As usual, we can write "accumulation = input − output." The accumulation of species z^A corresponds to the change in concentration in the void space of the system. This material balance gives

$$\frac{\beta}{W} C^A \frac{dz^A}{dt} = \left(\frac{F_F^A z_F^A}{W} + v_{-1} z^{Al}\right) - \left(\frac{F^A z^A}{W} + v_{+1} z^A\right) \quad (4.8)$$

Accumulation　　　　Input　　　　　　Output

Similar material balances can be written for Al and B. Note that in the case of Al there is no feed input or product output since the catalyst is in the form of a fixed bed in the stirred or recirculating reactor. The following additional equations are obtained:

$$C^{Al} \frac{dz^{Al}}{dt} = (v_{+1} z^A + v_{-2} z^B) - (v_{-1} z^{Al} + v_{+2} z^{Al}) \quad (4.9)$$

$$\frac{\beta}{W} C^B \frac{dz^B}{dt} = \left(\frac{F_F^B z_F^B}{W} + v_{+2} z^{Al}\right) - \left(\frac{F^B z^B}{W} + v_{-2} z^B\right) \quad (4.10)$$

where F_F^i is the inlet flow rate of feed species i ($i =$ A, B) (ml/sec); F^i, the outlet flow rate of product i ($i =$ A, B) (ml/sec); z_F^i, the fraction of tracer atoms of total atom in i ($i =$ A, B) in feed; z, the fraction of tracer atoms of total atoms in i ($i =$ A, B) in product; W, the weight of catalyst in the system (g); β, the volume of dead space (including that in catalyst pores, void space, and apparatus) (ml); C^i, the concentration of species i ($i =$ A, B) in gas-phase product (outlet) (ml/ml); C^{Al}, the concentration of species Al on the solid catalyst (ml/g); and t, the time (sec).

Equations (4.8)–(4.10) may be arranged as follows to show the coefficients corresponding to the variables z^A, z^{Al}, and z^B:

$$\frac{dz^A}{dt} = -\frac{W}{\beta C^A} \left(\frac{F^A}{W} + v_{+1}\right) z^A + \frac{W}{\beta} \frac{v_{-1}}{C^A} z^{Al} + \frac{F_F^A z_F^A}{\beta C^A} \quad (4.11)$$

$$\frac{dz^{Al}}{dt} = \frac{v_{+1}}{C^{Al}} z^A - \frac{v_{-1} + v_{+2}}{C^{Al}} z^{Al} + \frac{v_{-2}}{C^{Al}} z^B \quad (4.12)$$

$$\frac{dz^B}{dt} = \frac{W}{\beta} \frac{v_{+2}}{C^B} z^{Al} - \frac{W}{\beta C^B} \left(\frac{F^B}{W} + v_{-2}\right) z^B + \frac{F_F^B z_F^B}{\beta C^B} \quad (4.13)$$

In matrix form these equations may be written as

$$\frac{d}{dt}\begin{bmatrix} z_p^A \\ z^{Al} \\ z^B \end{bmatrix} = \begin{bmatrix} -\frac{W}{\beta C^A}\left(\frac{F^A}{W}+v_{+1}\right) & +\frac{W}{\beta}\frac{v_{-1}}{C^A} & 0 \\ \frac{v_{+1}}{C^{Al}} & -\left(\frac{v_{-1}+v_{+2}}{C^{Al}}\right) & +\frac{v_{-2}}{C^{Al}} \\ 0 & +\frac{W}{\beta}\frac{v_{+2}}{C^B} & -\frac{W}{\beta C^B}\left(\frac{F^B}{W}+v_{-2}\right) \end{bmatrix}$$

$$\times \begin{bmatrix} z^A \\ z^{Al} \\ z^B \end{bmatrix} + \begin{bmatrix} \frac{F_F^A z_F^A}{\beta C^A} \\ 0 \\ \frac{F_F^B z_F^B}{\beta C^B} \end{bmatrix} \quad (4.14)$$

Note that F^A, F_F^A, F^B, and F_F^B are known constants since tracing is superposed on a known steady-state reaction. For the simple system given by Eq. (4.14) the overall velocity of the reaction is given by

$$V = (F_F^A/W) - (F^A/W) \quad (4.15)$$

$$V = (F^B/W) - (F_F^B/W) \quad (4.16)$$

In addition, we know that

$$V = v_{+1} - v_{-1} = v_{+2} - v_{-2} \quad (4.17)$$

The concentrations C^A and C^B are also constant since the system operates at constant pressure and temperature. They can be computed from the feed rates of A and B to the system. The unknown parameters are the surface concentration C^{Al} and the unidirectional step velocity components, which may be taken as v_{+1} and v_{+2}. The observed transients in concentration of z^A and z^B together with the variable that is usually not observed z^{Al} constitute the dependent variables.

Even for the simple three-compartmental model represented by Eq. (4.15), it is not possible to obtain a general analytical solution in terms of the dependent variables z^A, z^{Al}, and z^B. Such a solution would require explicit expressions for the solution of a cubic equation in which a zero root does not occur. Computer solutions to this and more complicated models are discussed in Chapter 7. Generally the procedure is straightforward.

To illustrate the type of transient response data obtained in systems of this type, the following example employs a further simplification in which a series

4.5 Superposition Modeling with the Gradientless Reactor

of two unidirectional steps is assumed instead of both steps being taken as reversible as shown in Figure 4.1.

Example 4.5.1 Three Compartment Unidirectional Mechanism with a Recirculating Reactor

Consider the system represented by Eq. (4.7) in which a single unidirectional velocity V exists and the feed is pure A. Mass balances for the system are given by Eq. (4.8) and (4.9) with $v_{-1} = v_{-2} = 0$, $v_{+1} = v_{+2} = V$, and $F_F^B = 0$. Boundary conditions are assumed to be

$$z^A(0) = z^{Al}(0) = z^B(0) = 0, \qquad z_F^A(t) = z_0^A u(t)$$

where $u(t)$ is the unit step function. At $t = 0$, $u(t)$ changes from 0 to 1. These equations can be solved by standard methods for the three dependent variables z^A, z^{Al}, and z^B. The following solution is obtained:

$$z^A = z_0^A(1 - e^{-t/T_1})u(t)$$

$$z^{Al} = z_0^A \left(1 + \frac{T_1}{T_2 - T_1} e^{-t/T_1} + \frac{T_2}{T_1 - T_2} e^{-t/T_2} \right) u(t)$$

$$z^B = z_0^A \left[1 - \frac{T_1^2}{(T_1 - T_2)(T_1 - T_3)} e^{-t/T_1} - \frac{T_2^2}{(T_2 - T_1)(T_2 - T_3)} e^{-t/T_2} \right.$$
$$\left. - \frac{T_3^2}{(T_3 - T_1)(T_3 - T_2)} e^{-t/T_3} \right] u(t)$$

where

$$T_1 = \frac{\beta C^A}{VW + F^A}, \qquad T_2 = \frac{C^{Al}}{V}, \qquad T_3 = \frac{\beta C^B}{VW}$$

Since T_1 and T_3 contain no unknown constants they can be computed directly. In order to determine C^{Al}, it is necessary to observe z^B, assuming that z^{Al} is not observed. The problem is the inverse of solving for z^B with a known value of C^{Al}. Estimation of C^{Al} proceeds by a curve fitting technique. In order for C^{Al} to be estimated accurately, it is desirable for the surface concentration to be large in comparison with the concentrations C^A and C^B in the vapor phase. It is difficult in a circulating system to maintain a small dead volume β with respect to the amount of catalyst W. Thus, to maintain T_2 larger than $T_1 + T_2$, it is convenient to dilute the feed with an inert gas such as helium.

For purposes of illustration, let us assume that 100 ml/min of helium is used to dilute a feed $F_F^A = 1.0$ ml/min, and that the weight of catalyst $W = 1.0$ g. Then $F^A = 0.9$ ml/min, if the velocity $V = 0.1$ ml/g/min; $C^A = 0.009$ ml/ml and $C^B = 0.001$ ml/ml. If we take $C^{Al} = 0.2$ ml/g, the following values can be computed for the time constants with $\beta = 100$ ml:

$$T_1 = 0.891 \text{ min} = 53.5 \text{ sec}$$
$$T_2 = 2 \text{ min} = 120 \text{ sec}$$
$$T_3 = 0.991 \text{ min} = 59.4 \text{ sec}$$

Values of transient tracer concentrations are plotted in Figure 4.2. In practice a computer routine would be employed to determine the values of T_1, T_2, and T_3 from experimental data and from this the value of C^{Al} could be estimated.

Equation (4.7) represents in a general way the type of material balances that are obtained for a system with any number of compartments in which a transient for atomic tracer transfer is superposed on an overall steady-state mechanism. It is simply necessary to construct material balances for all the species in the same manner as is done for the simple three-compartment system illustrated by Eq. (4.14). It is possible to model systems in which mechanistic steps occur both in the forward and reverse directions without requiring any additional information about the nature of kinetics of individual reaction steps. Tracer transfer need not follow all steps of the mechanism that are being traced, since all atomic species may not be involved in each of the molecular species consumed and produced as in the case for an isomerization reaction. Independent data are obtained when different tracers follow the same reaction steps.

The number of compartmental models may exceed those developed by procedures similar to that discussed in Chapter 3 because exchange of isotope

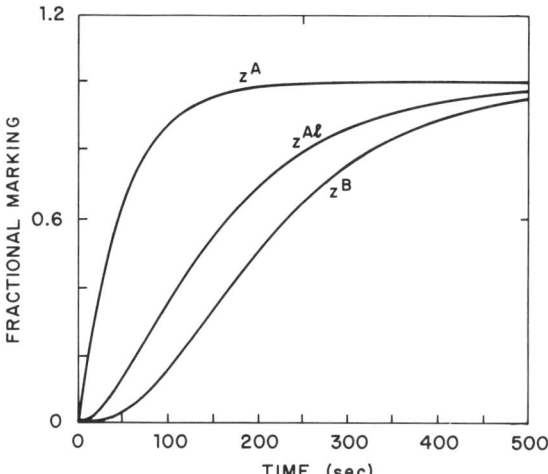

FIG. 4.2 Tracer transient for gradientless reactor, Example 4.5.1.

can occur with elements present in the catalyst or as inactive chemisorbed species that do not enter the steady-state mechanisms. Such contributions vanish when steady-state marking is reached after a period of time. Similarly exchanges can occur between various species that do not correspond to the chemical reaction being traced. It is, of course, also possible to model reactions in which the chemical reactions involved are at equilibrium by observing tracer redistribution.

Note that in the simple model illustrating Eq. (4.7), the **A** matrix and **b** vector are both taken as constants. If a step function is employed for marking the feed concentrations, z_F^A and z_F^B will be constant. This is the experimental procedure that we have employed with the additional restriction that only one terminal species is marked. To apply a pulse in a circulating system, it would be necessary to rapidly replace the entire volume of the circulating gaseous phase by one marked with one or more of the terminal species. In that case, **b** would be taken as zero. Usually we have arranged conditions so that the model involves **A** as a constant. This effects considerable simplification of the system equations solution, which then consists of a set of linear differential equations with constant coefficients. In some cases it has proved useful to develop models in which one or more of the state variables vary with time and in that case the **A** matrix will also be time dependent. The more general expression corresponding to such problems is

$$d\mathbf{z}/dt = p(t)\mathbf{z} + g(t) \tag{4.18}$$

where $p(t)$ and $g(t)$ are continuous over the range $\alpha < t < \beta$.

An important simplified class of modeling problems occurs when the tracer observations correspond to a steady-state regime. In that case $d\mathbf{z}/dt$ in Eq. (4.7) is zero and the system reduces to a set of simultaneous linear equations. It is then not possible to obtain information about the concentrations of intermediates C^{il} since the tracer concentration does not change during an experiment. The parameter groupings that can be determined will furnish information on step velocities, but often it is not possible to determine them completely. Chapter 5 is devoted to modeling tracer data obtained under such steady-state conditions.

As discussed previously, we do not address the problem of predicting the exact conditions for which heat and mass transfer effects are negligible compared with the rate of the intrinsic chemical reactions that are occurring. This subject has been considered at some length in a treatise by Aris (1975). Since the model equations employed in our studies are already complicated, it is desirable to operate under conditions where such effects do not occur. A simple way of ensuring absence of internal gradients is to conduct experiments with a range of particle sizes. It is then possible to determine directly what conditions should be employed so that particle size does not influence the overall reaction rate. Generally, it is possible to operate at a sufficiently low temperature so that such effects are not important. Because of the very much larger activation energies usually involved in chemical reactions as compared with diffusion, small changes in temperature will exert a large effect on the relative rates. The choice of recirculation rate in such systems has also been discussed by Nystrom (1978), employing published criteria for intraparticle transfer effects.

The employment of a recirculating reactor with high recirculation rates generally minimizes gradients of concentration and temperature external to the catalyst particles. As has been pointed out (Hudgins, 1981), the literature on chemical reactor engineering includes a number of diagnostic criteria for eliminating interphase transport-disguised kinetics. Hudgins shows that such criteria are inadequate because of the relatively low velocities employed in plug-flow laboratory reactors and suggests that the detection of intraphase transport intrusion must be made by examining measurements of reaction rate as a function of flow rate. It is possible to make such measurements readily in recirculating reactors when necessary.

A recirculating reactor requires a pump for moving gases, and this adds considerably to the dead volume compared with the volume occupied by the catalyst. Thus, the relative amount of terminal species contained in the gas phase to that adsorbed on the catalyst can become large, reducing sensitivity for determination of the chemisorbed concentration of such species.

Modifications of Eq. (4.7) can be used to model constant volume and plug-flow reactor systems, as will be discussed in the following sections. Such systems are generally not as convenient for the conducting of transient experiments. Not only must the reactions be conducted under differential conversion, but the modeling equations may become more complicated than in the case of gradientless reactor systems.

4.6 Superposition Modeling with the Constant Volume Reactor

In order to conserve consumption of expensive isotopes or when studying a very slow reaction, it is sometimes convenient to employ systems in which the feed is not continuously introduced and product withdrawn. Let us again consider the simple reaction system given by Eq. (4.14) as applied to such a closed system, in which the contents are still well mixed. In this case the concentrations of intermediates will change.

We assume that the kinetics of the overall reaction rate is not known and it is therefore necessary to operate so that conversion occurs at a nearly constant concentration of A for the period of time under observation. Such a mode of operation is described as differential and is often used in chemical

4.6 Superposition Modeling with the Constant Volume Reactor

kinetics experiments in which overall reaction rates are determined. In tracer superposition on a steady-state overall reaction it is also necessary to consider the conditions under which such operation will yield the desired parameters.

For a closed system operating in the regime of a constant conversion rate the C^i can be expressed as

$$C^B = C_0^A - \frac{WV}{\beta}t \qquad (4.19)$$

$$C^B = C_0^B + \frac{WV}{\beta}t \qquad (4.20)$$

A low conversion rate corresponds to $WV \to 0$ or to operation of the system under differential conditions. If C_0^A and C_0^B are both present in substantial concentrations, it may be possible to consider C^A and C^B as constants. In that case, in place of Eqs. (4.7) and (4.15), we have the following:

$$dz/dt = \mathbf{Az} \qquad (4.21)$$

or

$$\frac{d}{dt}\begin{bmatrix} z^A \\ z^{Al} \\ z^B \end{bmatrix} = \begin{bmatrix} -\frac{W}{\beta C^A}v_{+1} & +\frac{W}{\beta C^A} & 0 \\ \frac{v_{+1}}{C^{Al}} & -\frac{(v_{-1}+v_{+2})}{C^{Al}} & +\frac{v_{+2}}{C^{Al}} \\ 0 & \frac{W}{\beta C^B}v_{+2} & -\frac{W}{\beta C^B}v_{-2} \end{bmatrix}\begin{bmatrix} z^A \\ z^{Al} \\ z^B \end{bmatrix} \qquad (4.22)$$

which is the same form as for pulse introduction of tracer into an open system. If the C^i cannot be taken as constant a more complicated form results in which V might still be taken constant at appropriate values of WV

$$d(\mathbf{C}z)/dt = \mathbf{Az} \qquad (4.23)$$

An analytical solution can be readily obtained for Eq. (4.21). It is often the case in closed systems that β is large, and in that case when $W/\beta \ll 1$ the solution reduces from a three- to a two-compartmental model, yielding results equivalent to those obtainable by steady-state operation in a gradientless system. Such tracing will yield only the overall forward and reverse reaction velocities for the system. However, this mode of operation may be useful for obtaining data with less effort; because reaction conditions vary through a set of quasi-steady states as compared with the constant rates obtained for a single experiment with an open recirculating system.

Another special case of interest is obtained when a reaction is traced in a closed system that is at equilibrium. In that case the step velocities v_{+i} are equal and we have in place of Eq. (4.22)

$$\frac{d}{dt}\begin{bmatrix} z^A \\ z^{Al} \\ z^B \end{bmatrix} = \begin{bmatrix} -\frac{W}{\beta C^A}v_1 & +\frac{W}{\beta C^A}v_1 & 0 \\ \frac{v_1}{C^{Al}} & -\frac{(v_1+v_2)}{C^{Al}} & +\frac{v_2}{C^{Al}} \\ 0 & \frac{W}{\beta}\frac{v_2}{C^B} & -\frac{W}{\beta}\frac{v_2}{C^B} \end{bmatrix}\begin{bmatrix} z^A \\ z^{Al} \\ z^B \end{bmatrix} \quad (4.24)$$

An explicit solution is also possible for this system.

In studies involving closed systems, the introduction of a tracer would require the experimental technique of changing the gas-phase volume rapidly from one in which a tracer is absent to one including the tracer, unless a radioactive isotope is employed. In that case, introduction of a small concentration of marked species can be employed, and linearization is accomplished by a small perturbation of the system rather than by superposition.

4.7 Superposition Modeling with the Plug-Flow Reactor

A plug-flow reactor possesses the advantage that the catalyst volume can be maintained large compared with external void space. Also switching to tracer is readily accomplished without the need for purging of reactor volume dead space. As in the case of the closed system, it is necessary to operate a plug-flow reactor under differential conditions. Let us consider the three-compartmental model system discussed in Section 4.5 to illustrate the development of model equations for this type of system, which involves space as well as time as an independent variable. Figure 4.3 illustrates such a reactor, assumed to operate under isothermal and isobaric conditions. Note that this is not a compartmental model since a space variable is involved. The same reaction mechanism is, however, assumed as previously in Example 4.5.1.

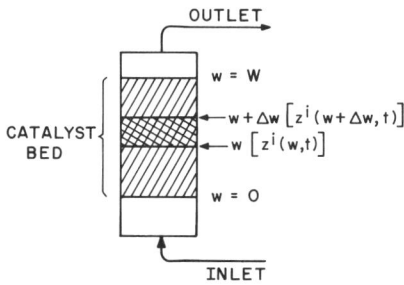

FIG. 4.3 Unidirectional plug-flow reactor.

4.7 Superposition Modeling with the Plug-Flow Reactor

It is assumed that no axial mixing occurs in the reactor. Reacting gases enter the catalyst bed at $w = 0$ and leave at $w = W$. Material balances over a differential reaction section give the following equations:

$$\frac{\partial F^A z^A}{\partial w} + \frac{\beta C^A}{W}\frac{\partial z^A}{\partial t} = v_{-1} z^{Al} - v_{+1} z^A \tag{4.25}$$

$$\frac{\partial F^B z^B}{\partial w} + \frac{\beta C^B}{W}\frac{\partial z^B}{\partial t} = v_{+2} z^{Al} - v_{-2} z^B \tag{4.26}$$

$$C^{Al}\frac{\partial z^{Al}}{\partial t} = v_{+1} z^A - v_{-1} z^{Al} - v_{+2} z^{Al} + v_{-2} z^B \tag{4.27}$$

These equations are somewhat similar to Eqs. (4.8)–(4.10). However, as in the case of the closed system we have the derivative of the product of two variables

$$\partial F^i z^i / \partial w = F^i(\partial z^i / \partial w) + z^i(\partial F^i / \partial w) \tag{4.28}$$

and furthermore, in addition to the space variable, the time derivatives of z^i are still present in the equations. For a differential reactor, we can take

$$F^A = F_0^A - Vw \tag{4.29}$$

$$F^B = F_0^B + Vw \tag{4.30}$$

but the terms in $\partial z^i / \partial w$ still remain. If the feed is pure component A, it is not possible to neglect the Vw terms in contrast to the simplification introduced in Eq. (4.23) for a closed volume reactor at quasi-steady state.

The following example, which parallels Example 4.5.1, illustrates the difference between compartmental and plug-flow modeling.

Example 4.7.1 Simplest Case for Plug-Flow Reaction

We wish to consider the same system as that represented by Example 4.5.1, in which a single unidirectional velocity exists. The mass balances for the system are given by Eqs. (4.25) and (4.27), in which there are the further restrictions that $v_{-1} = v_{-2} = 0$, $v_{+1} = v_{+2} = V$, and $F_F^B = 0$. The boundary conditions are assumed to be

$$Z^A(0, w) = 0, \quad \forall w > 0$$

$$Z^B(0, w) = 0, \quad \forall w > 0$$

$$Z^{Al}(0, w) = 0, \quad \forall w > 0$$

$$Z^A(t, 0) = Z_0^A u(t)$$

$$Z^B(t, 0) = 0$$

where $u(t)$ is the unit step function. The equations can be solved conveniently by use of the Laplace transform

$$z^A(t, w) = z_o^A u\left(t - \frac{\beta w}{WF}\right)$$

$$z^{Al}(t, w) = z^B(t, w) = z_o^A\left[1 - \exp\left(-\frac{V}{C^{Al}}\right)\left(t - \frac{\beta w}{WF}\right)\right]u\left(t - \frac{\beta w}{WF}\right)$$

where $u[t - (\beta w/FW)]$ is the unit step function that has a value of zero at time less than $\beta w/WF$ and unity at time greater than $\beta w/WF$ (w is the distance through the bed). It varies from zero to W, the total amount of catalyst. The measurable entities are Z^A and Z^B at the output of the reactor:

$$Z^A(t, W) = z_o^A u[t - (\beta/F)]$$

$$z^{Al}(t, W) = z^B(t, W) = z_o^A\left[1 - \exp\left(-\frac{V}{C^{Al}}\right)\left(t - \frac{\beta}{F}\right)\right]u\left(t - \frac{\beta}{F}\right)$$

Let us assume, for comparison with Example 4.5.1 that although the partial pressure in the plug-flow case is higher, because no helium diluent is needed, we will still take $V = 0.1$ ml/g/min.

It is necessary to operate the plug-flow system under differential conversion. Let us assume that 5% conversion will be approximately differential and that C^{Al} will still be the same as previously:

$$C^{Al} = 0.2 \text{ ml/g}$$

The time constant $C^{Al}/v = 2$ min $= 120$ sec, as in Example 4.5.1 assume that 10 g of catalyst would be needed for the plug-flow reactor. This would correspond to 1 ml/min of product, requiring a feed rate $F = 20$ ml/min. In the case of a plug-flow reactor the void volume, assuming that marked feed enters directly at the bed inlet, corresponds to just the space in the bed not occupied by solid catalyst. If we take this nominally at $\beta = 5$ ml for the 10 g of catalyst,

$$\beta/F = \tfrac{5}{20} \text{ min} = 15 \text{ sec}$$

Values of the tracer transients corresponding to these conditions are plotted in Fig. 4.4. The result when compared with Fig. 4.1 for the corresponding compartmental reactor shows that the separations are sharper. The Z^A transient is a step function in place of an exponential. The Z^B transient is an exponential that does not involve the void space β so C^{Al} can be more readily estimated than in a circulating system. Also, if the appropriate differential operation can be achieved, a plug-flow reactor could be operated

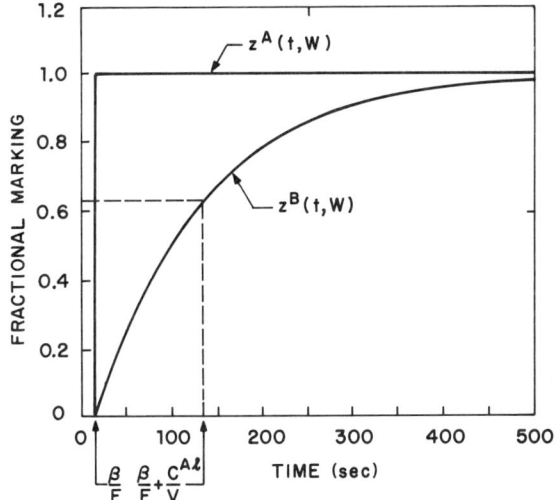

FIG. 4.4 Tracer transient for plug-flow reactor, Example 4.7.1.

at high pressure without requiring dilution of the feed necessary in a compartmental reactor.

Useful relationships for *steady-state* tracing in plug-flow reactors are given in Chapter 5, but in that case it is not possible to obtain the intermediate concentrations by tracing, as in the case of transient modeling.

It is important to note that even with differential conversion of reactants in a plug-flow reactor it is not simple to model the transient tracer operation for the case in which the intermediate velocities are not *unidirectional*, because v_{-1} and v_{-2} must also be taken as constants. At very high flow rates, tracer transfer as well as overall conversion would become differential.

Analogous equations for a transient reaction in a flow system are given by Bennett (1976). In that case the reaction itself is assumed to be transient as contrasted to the situation discussed here in which the reaction system is at constant velocity with a transient superposition of tracer. When a plug-flow system is operated at sufficiently high velocity the space derivatives in Eqs. (4.25)–(4.27) become negligible compared with the time derivatives. In the case of a recirculating reactor system the external input and output streams constitute contributions that replace these terms.

List of Symbols

A	Chemical substance
Al	Chemisorbed intermediate

A	Matrix of parameters
B	Chemical substance
b	Vector containing feed rates
C^i	Concentration of species i ($i = $ A, B) in gas-phase product for gradientless reactor (ml/ml)
C_0^i	Initial or inlet concentration of species i ($i - $ A, B)
C^{Al}	Concentration of species A on solid catalyst (ml/g)
Cz	Vector of $C^i z^i$ tracer contents ($i = $ A, A, B)
F_F^A	Inlet flow rate of A (ml/min)
F_F^B	Inlet flow rate of B (ml/min)
F^A	Outlet flow rate of A (ml/min)
F^B	Outlet flow rate of B (ml/min)
g	Vector of constants
g(t)	Vector of functions of time
l	Surface site
P(t)	Matrix of time dependent parameters
t	Time (min)
$u(t-k)$	Unit step function with value of 0 at $t < k$ and unity at $t > k$
V	Overall reaction velocity (ml/min/g)
$v_{\pm i}$	Mechanistic step velocities (ml/min/g)($i = 1, 2$)
W	Total weight of catalyst bed (g)
w	Degree of progress (distance) through bed (g)
x_i	Stable or untraced component $i = (1, 2)$
x_i^*	Tagged or traced component $i = (1, 2)$
z_F^A	Inlet fractional tracer marking of species A
z_F^B	Inlet fractional tracer marking of species B
z^i	Tracer marking in circulating stream or product of gradientless reactor system, $i = $ (A, Al, B)
$z^i(t, w)$	Time and space variant tracer marking for plug-flow system of a species, $i = $ (A, Al, B)
z	Vector of the state variables
β	Dead space including voids in catalyst bed (ml)
θ	Vector of the parameters
λ	Variable in Eq. (4.6)
$\forall$	For any

References

Aris, R. (1969). *Ind. Eng. Chem.* **61**, 17.
Aris, R. (1975). "The Mathematical Theory of Diffusion and Reaction in Permeable Catalysts," 2 volumes. Clarendon Press, Oxford.
Bard, Y. (1974) "Non-Linear Parameter Estimation." Academic Press, New York.

References

Bennett, C. O. (1976). *Catal. Rev. Sci. Eng.* **13**(2), 121.
Berman, M. (1971). Compartmental Modeling, *in* "Advances in Medicine Physics" (Laughlin and Webster, eds), pp 279–295. Second Int'l Congress on Medical Physics, Inc. Boston.
Berman, M. (1979). *Prog. Biochem. Pharmacol.* **15**, 92–126.
Biloen, P., Helle, M. N., and Sachtler, W. M. H. (1979). *J. Catal.* **58**, 95.
Brownell, G. L., Berman, M., and Robertson, J. S. (1968). *Int. J. Radiat. Isotopes* **19**, 249.
Enemoto, S., and Horiuti, J. (1953). *J. Res. Inst. Catal., Hokkaido Univ.* **2**, 87.
Happel, J. (1978). *Chem. Eng. Sci.* **33**, 1567.
Happel, J., and Kiang, S. (1977). *J. Res. Inst. Catal. Hokkaido Univ.* **25**, 99.
Happel, J., Suzuki, J., Kokayeff, P., and Fthenakis, V. (1980). *J. Catal.* **65**, 59.
Happel, J., Cheh, H. Y., Otarod, M., Ozawa, S., Severdia, A. G., Yoshida, T., and Fthenakis, V. (1982). *J. Catal.* **72**, 314.
Hudgins, R. R. (1981). *Chem. Eng. Sci.* **36**, 1579.
Kummer, J. T., de Witt, T. W., and Emmett, P. H. (1948). *J. Am. Chem. Soc.* **70**, 3632.
Le Cardinal, G. (1978). *Chem. Eng. Sci.* **33**, 1568.
Le Cardinal, G., Walter, E., Bertrand, P., Zoulalian, O., and Gelus, M. (1977). *Chem. Eng. Sci.* **32**, 733.
Nystrom, M. G. (1978). *Chem. Eng. Sci.* **33**, 390.
Otarod, M., Ozawa, S., Yin, F., Chew, M., Cheh, H. Y., and Happel, J. (1983). *J. Catal.* **84**, 156.
Ponec, V., and van Barneveld, W. A. (1979). *Ind. Eng. Chem. Prod. Res. Div.* **18**, 268.
Rice, O. K. (1960). *J. Am. Chem. Soc.* **65**, 1851.
Schulz, H., and Achtsnit, H. D. (1976). *Proc. Iberoameric. Congr. Catal., Vth, Lisbon.*
Temkin, M. I. (1979). *Adv. Catal.* **28**, 173.
Walter, E. (1975). These de 3°Cycle-Universite Paris XI.
Yang, K. H., and Hougen, O. A. (1950). *Chem. Eng. Prog.* **46**, 146.

Chapter 5

Steady-State Tracing

5.1 Introduction

As was noted in the previous chapter, the steady-state tracing equations are a special case of the general equations for superposition modeling in which the time derivatives become zero. In steady-state tracing we are not able to determine surface concentrations of intermediates because the concentrations of tracer in such species do not change in the course of an experiment. The parameters to be determined are the velocities of individual mechanistic steps. An earlier survey (Happel, 1972) discusses some of the general principles involved and gives a number of industrial catalysis examples. At about the same time Emmett (1972) published a general review of the use of isotopic tracers in catalysis that was devoted to steady-state tracing. His review for the most part is not devoted to modeling for the purpose of estimating rates of individual reaction steps.

While less comprehensive than transient modeling, the steady-state method can be applied more readily in the case of plug-flow reactors because of the elimination of the time variable present in transient tracing. Steady-state tracing is also advantageous where exchange of tracer atoms can occur with the catalyst support or with a molten phase in the catalyst pores, because of possible difficulties in the interpretation of transient data.

In an elegant statistical-mechanical development, Horiuti showed how transition-state theory could be applied to developing a relationship between forward and reverse velocities of individual reaction steps in a mechanism. These ideas were further developed and summarized by Horiuti and

Nakamura (1967) and Horiuti (1973). Such relationships are especially useful in steady-state tracing because the information obtained by the tracing procedure itself is more limited than in the case of transient tracing. The ideas are detailed in the following sections of this chapter. Before considering them we wish to examine the general problem of parameter estimation from information obtainable by observation of tracer levels in terminal compartments with steady-state tracing.

Often it is convenient to employ a gradientless recirculating reactor system as in the case of transient tracing. A tracer experiment then consists of setting tracer concentrations at fixed levels in the feed for one or more components and observing the corresponding concentrations in the product stream. The transfer rate of tracer to the traced species in the gas phase is then directly calculated as (Happel and Hnatow, 1976)

$$-a_e^i = m^i\left(\frac{F_F^i z_{Fe}^i}{W} - \frac{F^i z_e^i}{W}\right) \tag{5.1}$$

where a_e^i is the transfer rate of traced element e to species i ($g_{atoms}/sec/g_{cat}$), F^i, the rate of removal of species i from the reactor (moles/sec), z_{Fe}^i, the fraction of tracer element e in feed of component i, which is marked, z_e^i the fraction of tracer element e in effluent of component i, which is marked, W, the weight of catalyst (g); and m^i, the number of atoms of tracer element in a molecule of i. The subscript e can be omitted when only a single element is used as a tracer.

In the case of a fixed-bed plug-flow reactor, where integration must be conducted for flow through the reactor

$$a_e^i = m^i \frac{d(F^i z_e^i)}{dw} \tag{5.2}$$

When degree of conversion and tracer transfer are both differentials the derivative in Eq. (5.2) can be replaced by $\Delta(F^i z_e^i)/W$.

In order to estimate the velocities or velocity groupings of intermediate reaction steps in an assumed mechanism, sets of a^i and z^i can be obtained corresponding to given marking levels of tracers introduced into a reaction system, all such experiments being conducted at constant reaction conditions, including a constant overall velocity. These observed quantities are related to velocity groupings by material balances constructed for each of the terminal species n and intermediates k, as discussed in the previous chapter. One material balance corresponds to the condition of no change in the number of tracer atoms in the course of passage through the reactor

$$\sum_{i=1}^{i=n} a_e^i = 0$$

Thus, there are $k + n - 1$ independent balances corresponding to input and output of tracer to each species less one. The next step is to eliminate the markings of tracer in the intermediates, since it is assumed that they are not observed. Then there remain $n - 1$ equations, corresponding to the possibility of determining $n - 1$ parameters. These equations consist of linear relationships expressing velocity groupings in terms of observed transfer rates. A unique solution can be obtained for such systems of equations (see Park and Himmelblau, 1982). The relationship between groupings and transfer data must be obtained under constant reaction conditions in which the step velocities $v_{\pm i}$ remain the same while tracer level is changed. Depending on the degree of branching, a number of such experiments can be conducted, each set corresponding to a specified reaction condition with constant v_{+i}. For a catenary system with no branching and only one parameter, the overall velocity ratio can be determined and regardless of the marking level each tracer experiment will enable it to be computed. Additional branches in a mechanism will correspond to the possibility of determining additional velocity groupings, but in order to do so additional tracer experiments at different marking levels will be required.

Velocity groupings obtained in this manner can then be broken down into similar groups, and sometimes into individual step velocities by making use of the relationships developed by Horiuti (discussed in Sections 5.2 and 5.3). These relationships, together with the considerations previously noted, are applied in Sections 5.4 and 5.5. Horiuti's relationships are also applicable in transient modeling and rate-equation construction.

5.2 Step-Velocity Grouping

When a reaction occurs at steady state, constant concentrations of intermediates exist on catalyst sites without any substantial passage of intermediates to the effluent stream. It is then possible to make simple material balances to determine how the overall net velocity (or the net velocity in any portion of a sequence) is related to the velocities of individual mechanistic steps (Happel, 1972).

Suppose that a given sequence is designated by steps numbered $1, 2, \ldots, n$ with corresponding mechanistic steps whose rates are given by $v_{+1}, v_{+2}, v_{+3}, \ldots, v_{+n}$. The rate of transformation by the first step v_{+1} can be expressed as occurrences per unit time per unit mass of catalyst. To obtain the corresponding rate of transformation for the overall reaction, it is necessary to divide v_{+1} by the stoichiometric number v (Horiuti, 1973). The stoichiometric number v_s for an elementary step represents the number of times that it occurs

5.2 Step Velocity Grouping

for each time that the molecular change, represented by the overall reaction, occurs once, as previously discussed in Chapter 3. Thus, the rate of transformation of species for the overall reaction in the forward direction for the first step is $V_+^1 = v_{+1}/v_1$. When we move to the second mechanistic step in the sequence, an intermediate produced by step 1 (which is a reactant for step 2) can now either proceed forward at velocity v_{+2} or revert to the original species at velocity V_-^1. If these probabilities are both referred to a single occurrence of the overall reaction, the probability of the forward transformation proceeding through a second step is

$$V_+^{1,2} = V_+^1 \frac{v_{+2}/v_2}{v_{+2}/v_2 + V_-^1} \tag{5.3}$$

Since $V_+^1 = v_{+1}/v_1$ and $V_-^1 = v_{-1}/v_1$, we may substitute these expressions in Eq. (5.3) and obtain

$$V_+^{1,2} = \frac{1}{(v_1/v_{+1}) + (v_{-1}v_2/v_{+1}v_{+2})} \tag{5.4}$$

At the same time $V_-^{1,2}$ is equal to $v_{-2}/v_2 \times$ the probability of the species going backward through step 1 or

$$V_-^{1,2} = \frac{v_{-2}}{v_2} \frac{V_-^1}{V_-^1 + (v_{+2}/v_2)} = \frac{1}{(v_2/v_{-2}) + (v_{+2}v_1/v_{-2}v_{-1})} \tag{5.5}$$

Such a sequence can be continued through the entire series of steps to final products, and a similar series can be developed for the overall reverse velocity. Thus, we finally obtain

$$V_+^{1,2,3,\ldots,n} = \frac{1}{\dfrac{v_1}{v_{+1}} + \dfrac{v_{-1}v_2}{v_{+1}v_{+2}} + \dfrac{v_{-1}v_{-2}v_3}{v_{+1}v_{+2}v_{+3}} + \cdots + \dfrac{v_{-1}v_{-2}\cdots v_{-(n-1)}v_n}{v_{+1}v_{+2}\cdots v_{+(n-1)}v_{+n}}} \tag{5.6}$$

$$V_-^{1,\ldots,(n-2),(n-1),n} = V_-^{1,2,3,\ldots,n}$$

$$= \frac{1}{\dfrac{v_n}{v_{-n}} + \dfrac{v_{+n}v_{(n-1)}}{v_{-n}v_{-(n-1)}} + \dfrac{v_{+n}v_{+(n-1)}v_{(n-2)}}{v_{-n}v_{-(n-1)}v_{-(n-2)}} + \cdots + \dfrac{v_{+n}v_{+(n-1)}\cdots v_{+2}v_1}{v_{-n}v_{-(n-1)}\cdots v_{-2}v_{-1}}} \tag{5.7}$$

where the v_i's represent the stoichiometric number of corresponding steps. If a marked species is introduced as a reactant with an unmarked product being present in the reaction system, the rate at which marked product species will appear is given by V_+ in Eq. (5.6). This may be considered a unidirectional forward velocity. Similarly, if an unmarked initial reactant in a series path were present in a system and if the product in the system were marked, the rate at which marked species of initial reactant were formed would be given by V_- in Eq. (5.7). This could then be considered as a reverse unidirectional velocity. Groupings of this type apply equally well to any portion of a chain of reaction steps, instead of only the chain connecting terminal products.

In measurement of reaction rates without the use of tracers, the velocity at which species are produced is measured by the *net* appearance or disappearance of species. In a chain of steps corresponding to a simple overall reaction the following relationship applies:

$$V = (v_{+1} - v_{-1})/v_1 = (v_{+2} - v_{-2})/v_2 = \cdots = (v_{+n} - v_{-n})/v_n \qquad (5.8)$$

where V is the overall reaction rate corresponding to the sequence. Temkin (1979) has shown how we can relate this velocity to the forward and reverse velocities of mechanistic steps. First, the identity may be written as

$$(v_{+1} - v_{-1})v_{+2}v_{+3}\cdots v_{+n} + v_{-1}(v_{+2} - v_{-2})v_{+3}\cdots v_{+n}$$
$$+ v_{-1}v_{-2}(v_{+3} - v_{-3})\cdots v_{+n} + v_{-1}v_{-2}v_{-3}\cdots(v_n - v_{-n})$$
$$= v_{+1}v_{+2}v_{+3}\cdots v_n - v_{-1}v_{-2}v_{-3}\cdots v_{-n} \qquad (5.9)$$

This identity can be easily verified by multiplying out the terms in parentheses. By using Eq. (5.8) we can then obtain

$$(v_1 v_{+2} v_{+3}\cdots v_{+n} + v_{-1} v_2 v_{+3}\cdots v_{+n} + v_{-1} v_{-2} v_3 \cdots v_{+n} + \cdots + v_{-1} v_{-2} v_{-3}\cdots v_n)V$$
$$= v_{+1} v_{+2} v_{+3}\cdots v_n - v_{-1} v_{-2} v_{-3}\cdots v_{-n} \qquad (5.10)$$

Comparison with the relationship for V given by Eq. (5.10) with the expressions for V_+ and V_- given by Eqs. (5.6) and (5.7) gives

$$V^{1,2,3,\ldots,n} = V_+^{1,2,3,\ldots,n} - V_-^{1,2,3,\ldots,n} \qquad (5.11)$$

Division of Eq. (5.6) by Eq. (5.7) provides

$$\frac{V_+^{1,2,3,\ldots,n}}{V_-^{1,2,3,\ldots,n}} = \frac{v_{+1} v_{+2} v_{+3}\cdots v_{+n}}{v_{-1} v_{-2} v_{-3}\cdots v_{-n}} \qquad (5.12)$$

Unless the forward and reverse velocities can be uniquely identified with a single unbroken path that includes all intermediates, the concept of a single overall forward and reverse velocity for a reaction is not unique. The identification of velocity groupings then becomes more complicated. For example, if a traced atomic species is distributed between two product species in a mechanistic step, two separate unidirectional rates of transfer can be written for transfer to each separate product species containing tracer.

When a multiple overall reaction occurs the relationships between step velocities are more complicated. It is, of course, always possible to measure rates of atomic transfer between any terminal species containing the same atom. These velocity groupings will be used following the next section to show results that can be obtained by steady-state tracing experiments.

5.3 Transition-State Theory and Thermodynamics

One of the important developments in physical chemistry is the notion that chemical reactions are not kinetically simple but proceed through a number of steps or stages, each of which is called an elementary reaction. The mechanisms that have been discussed in this book ideally would consist of sets of such steps. Although this is not always the case, it is of interest to examine the behavior of such mechanistic systems since the most satisfactory theories deal with the rates of such elementary steps. The quantitative formulation of absolute reaction rates, first extensively used in the work of Eyring (1935), was elaborated and applied to heterogeneous catalysis systems by Horiuti and Ikusima (1939) and Horiuti (1973). The form used here, applied to a set of elementary reactions comprising a mechanism, was discussed by Hollingsworth (1957).

The basic initial assumption employed, which links kinetics and thermodynamic behavior (Happel, 1980), is the following equation relating to a single elementary step:

$$v_{+i}/v_{-i} = \exp(-\Delta g_i/RT) \quad (5.13)$$

where $v_{\pm i}$ are the rates of conversion per unit mass of catalyst for each forward and reverse mechanistic step i; Δg_i, the Gibbs free energy change for step i; T, the absolute temperature; and R, the gas constant.

For a single overall reaction the free-energy change is the sum of those for each step multiplied by the stoichiometric number of the step:

$$\Delta G = \sum_{i=1}^{i=n} v_i \Delta g_i \quad (5.14)$$

Equations (5.13) and (5.14) may be combined to obtain

$$\prod_{i=1}^{i=n} \left(\frac{v_{+i}}{v_{-i}}\right)^{v_i} = \exp-\frac{\Delta G}{RT} \quad (5.15)$$

The following example will show how the concepts developed in this and the previous section can be applied to simple reactions.

Example 5.3.1

Consider the catalytic oxidation of sulfur dioxide to proceed by the mechanism shown in the following table.

Step	Reaction	Stoichiometric number
1	$O_2 + 2l \underset{v_{-1}}{\overset{v_{+1}}{\rightleftharpoons}} 2Ol$	1
2	$SO_2 + l \underset{v_{-2}}{\overset{v_{+2}}{\rightleftharpoons}} SO_2 l$	2
3	$SO_2 l + Ol \underset{v_{-3}}{\overset{v_{+3}}{\rightleftharpoons}} SO_3 l + l$	2
4	$SO_3 l \underset{v_{-4}}{\overset{v_{+4}}{\rightleftharpoons}} SO_3 + l$	2
	$2SO_2 + O_2 = 2SO_3$	

It is desired to obtain a relationship between step and path velocities and the overall equilibrium constant, assuming given partial pressures of reacting species.

The overall free energy change will be the sum of the free energy changes for each step multiplied by the number of times it is required to contribute to the material balance to produce the overall reaction. Thus,

$$\Delta G = \Delta g_1 + 2\Delta g_2 + 2\Delta g_3 + 2\Delta g_4$$

$$\Delta G = -RT \ln \left[K_p \frac{p_{SO_3}^2 p_{O_2}}{p_{SO_2}^2} \right]$$

and

$$\Delta G = -RT \ln \left(\frac{v_{+1}}{v_{-1}} \right) - 2RT \ln \left(\frac{v_{+2}}{v_{-2}} \right) - 2RT \ln \left(\frac{v_{+3}}{v_{-3}} \right) - 2RT \ln \left(\frac{v_{+4}}{v_{-4}} \right)$$

Combining these expressions, we obtain

$$\frac{K_p p_{SO_2}^2 p_{O_2}}{p_{SO_3}^2} = \frac{v_{+1}}{v_{-1}} \left[\frac{v_{+2} v_{+3} v_{+4}}{v_{-2} v_{-3} v_{-4}} \right]^2$$

If sulfur were employed as a tracer, it would be possible to determine the forward and reverse velocities in the path of atomic transfer comprising steps 2, 3, and 4. According to Eq. (5.12), we could write

$$K_p \frac{p_{SO_2}^2 p_{O_2}}{p_{SO_3}^2} = \frac{v_{+1}}{v_{-1}} \left[\frac{V_+^{2,3,4}}{V_-^{2,3,4}} \right]^2$$

Developments like this can be employed in the construction of rate expressions, as will be shown in Chapter 8. In the modeling of mechanisms they furnish useful additional relationships between velocity groupings and thermodynamics.

In the case of multiple overall reactions, it is possible to express a given mechanism in terms of component simple reactions among terminal species. Any given experiment will correspond to a set of step velocities together with reaction extents of an arbitrarily chosen set of consistent overall reactions. If the step velocities are apportioned to each of the overall reactions, relationships similar to Eq. (5.15) can be written for each of them.

5.4 Single-Path Reactions

We will use the information thus far developed to show how some parameters can be obtained by steady-state tracing. Parodoxically, steady-state tracing will not enable individual step velocities to be determined for the simplest case in which no branching of reaction steps occurs. Consider the overall reaction

$$A = B \tag{5.16}$$

with the reaction mechanism

$$A + l \underset{v_{-1}}{\overset{v_{+1}}{\rightleftharpoons}} Al$$

$$Al \underset{v_{-2}}{\overset{v_{+2}}{\rightleftharpoons}} B + l \tag{5.17}$$

where l is a catalyst site and Al is a chemisorbed species. Let us suppose that A and B each contain only a single atom of traced element. There are $n = 2$ species and $k = 1$ intermediates involved. Hence, there will be $n + k - 1 = 2$ material balance equations for a recirculating reactor system:

$$-a^A = z^A v_{+1} - z^{Al} v_{-1}, \qquad -a^A = a^{Al} v_{+2} - z^B v_{-2} \tag{5.18}$$

Elimination of z^{Al} gives

$$-a^A = \frac{z^A v_{+1} v_{+2}}{v_{-1} + v_{-2}} - \frac{z^B v_{-1} v_{-2}}{v_{-1} + v_{+2}} = V_+^{1,2} z^A - V_-^{1,2} z^B \tag{5.19}$$

Since $V = V_+ - V_-$, we can eliminate either V_+ or V_- from Eq. (5.19) to obtain

$$V_+^{1,2} = (a^A + Vz^B)/(z^B - z^A) \tag{5.20}$$

or

$$V_-^{1,2} = (a^A + Vz^A)/(z^B - z^A) \tag{5.21}$$

Division of Eq. (5.20) by Eq. (5.21) gives a relationship for the velocity ratio

$$\frac{V_+^{1,2}}{V_-^{1,2}} = \frac{a^A + Vz^B}{a^A + Vz^A} = \frac{v_{+1}v_{+2}}{v_{-1}v_{-2}} \qquad (5.22)$$

Tracing will not enable the four unknowns $v_{+1,2}$ to be determined.

It is sometimes useful to calculate the velocity ratio $V_+^{1,2}/V_-^{1,2}$ entirely from tracer levels, without using a determination of V. This can be accomplished for a recirculating reactor system by making use of the fact that since the velocity remains constant

$$F^A = F_F^A - VW \qquad (5.23)$$

From Eqs. (5.16) and (5.19)

$$(F_F^A z_F^A/W) - (F^A z^A/W) = z^A V_+ - z^B V_- \qquad (5.24)$$

Using Eq. (5.23) to eliminate F^A, we obtain

$$V_- = \left(\frac{F_F^A}{W}\right)\left(\frac{z_F^A - z^A}{z^A - z^B}\right) \qquad (5.25)$$

Similarly, we have

$$F^B = F_F^B + VW \qquad (5.26)$$

and can derive in the same manner

$$V_+^{1,2} = \left(\frac{F_F^B}{W}\right)\left(\frac{z^B - z_F^B}{z^A - z^B}\right) \qquad (5.27)$$

Division of Eq. (5.27) by Eq. (5.25) gives

$$\frac{V_+^{1,2}}{V_-^{1,2}} = \left(\frac{F_F^B}{F_F^A}\right)\left(\frac{z^B - z_F^B}{z_F^A - z^A}\right) \qquad (5.28)$$

It is readily shown for any reaction governed by Eq. (5.16) that follows a sequence

$$A \underset{v_{-1}}{\overset{v_{+1}}{\rightleftharpoons}} X'l \cdots X''l \underset{v_{-n}}{\overset{v_{+n}}{\rightleftharpoons}} B \qquad (5.29)$$

General relationships for recirculating reactors can be obtained similar to those derived for a two-step sequence. Thus, for example, instead of Eq. (5.28), we can write

$$\frac{V_+^{1,2,\ldots,n}}{V_-^{1,2,\ldots,n}} = \frac{v_{+1}v_{+2}\cdots v_{+n}}{v_{-1}v_{-2}\cdots v_{-n}} = \frac{F_F^B}{F_F^A}\left[\frac{z^B - z_F^B}{z_F^A - z^A}\right] \qquad (5.30)$$

The same equation will also apply to the transfer of tracer that follows a single

5.4 Single-Path Reactions

unbroken path even if untraced species form separate branches. If all rate-controlling steps are in the path of the tracer atoms in such a sequence, Eq. (5.15) will also apply.

While the use of an open recirculating reactor is desirable for transient tracing, its use for steady-state tracing is not necessary. In steady-state tracing the use of a plug-flow reactor introduces only the space variable, so it is possible to integrate the tracer transfer equations without the requirement for knowledge of chemisorption kinetics (discussed in Section 4.7), provided that the overall conversion, is differential. In that case, the amount of tracer transfer may still be larger, so a_i is a differential, as given by Eq. 5.2. In the following paragraphs a derivation for steady-state, plug-flow tracing is given. We will again assume that $m^i = 1$, so both A and B contain only a single atom of traced species; from Eq. (5.2) we have

$$-d(F^A z^A)/dw = d(F^B z^B)/dw = z^A V_+^{1,2,\ldots,n} - z^B V_-^{1,2,\ldots,n} \quad (5.31)$$

The first of these derivatives is expanded as follows

$$\frac{-d(F^A z^A)}{dw} = -F^A \frac{dz^A}{dw} - z^A \frac{dF^A}{dw} \quad (5.32)$$

However, for an overall differential reaction, the overall velocity V is constant and equal to

$$-dF^A/dw = dF^B/dw = V \quad (5.33)$$

From Eqs. (5.31)–(5.33), we obtain

$$(-F^A \, dz^A/dw) = (z^A - z^B) V_-^{1,2,\ldots,n} \quad (5.34)$$

Since tracer is not destroyed, a material balance enables z^B to be expressed as

$$z^B = (F_F^A z_F^A + F_F^B z_F^B - F^A z^A)/F^B \quad (5.35)$$

Substitution of this value of z^B into Eq. (5.34) and rearranging for integration gives

$$\int_{z_F^A}^{z_W^A} \frac{dz^A}{z^A - (F_F^A z_F^A + F_F^B z_F^B / F^A + F^B)} = -V_-^{1,2,\ldots,n}(F^A + F^B) \int_0^W \frac{dw}{F^A F^B} \quad (5.36)$$

$F^A + F^B$ is constant since the total moles do not change. Analogous to Eq. (5.34), we may write for tracer in species B:

$$F^B \frac{dz^B}{dw} = (z^A - z^B) V_+^{1,2,\ldots,n} \quad (5.37)$$

which may be expressed in a form similar to Eq. (5.36)

$$\int_{z_F^B}^{z_W^B} \frac{dz^B}{(F_F^A z_F^A + F_F^B z_F^B/F^A + F^B) - z^B} = -V_+^{1,2,\ldots,n}(F^A + F^B) \int_0^W \frac{dw}{F^A F^B}$$

(5.38)

By integrating the left-hand side of Eqs. (5.36) and (5.38) and combining so as to eliminate the variable w, we obtain

$$\frac{V_+^{1,2,\ldots,n}}{V_-^{1,2,\ldots,n}} = \ln\left(\frac{z_F^B - z_M}{z_F^B - z_M}\right) \bigg/ \ln\left(\frac{z_W^A - z_M}{z_F^A - z_M}\right)$$

(5.39)

where Z_M is the average z of the mixture, which remains constant

$$z_M = \frac{F_F^A z_F^A + F_F^B z_F^B}{F^A + F^B} = \frac{F_W^A z_W^A + F_W^B z_W^B}{F^A + F^B}$$

(5.40)

Note that although V is assumed constant, it is not possible to take F^A and F^B at constant average values in performing the integrations as is the case when correlating overall rate data for differential plug-flow reactors. The reactor is differential in the sense that partial pressure changes must be small enough so that the velocity remains essentially constant in both directions. This would not be the case, for example, if an experiment were conducted for dehydrogenation of pure butane with only a small degree of conversion because the partial pressure of hydrogen produced would vary considerably causing V_- to vary from reactor inlet to outlet.

It can be shown that for differential rate of tracer transfer and constant overall velocity, Eq. (5.39) for a plug-flow reactor reduces to Eq. (5.30). In that case, the marking difference between A and B will be small.

The following example illustrates the application of these equations to a plug-flow reactor system based on studies reported by Happel *et al.* (1973a).

Example 5.4.1

The mechanism of the isobutane–isobutene–hydrogen system over a commercial chromia–alumina (nominally containing 20% wt. of chromia; supplied by the Houdry Process Corp.) catalyst was studied by Kamholz (1970) in a differential reactor using radioactive carbon, ^{14}C-marked isobutane, and isobutene. For a typical experiment (Run 85-1) the feed composition and tracer markings shown in the following table were reported. We will employ Eq. (5.39) to calculate V_+/V_- in the path of carbon transfer. Since radioactive counts are directly proportional to the marking level, instead of attempting to calculate the actual marked fraction (which is very small) we will simply denote markings by z' employing these quantities in the relevant equations. Also since the number of atoms marked in isobutane and isobutene

5.4 Single-Path Reactions

is the same, it is not necessary to modify Eq. (5.40) for use in this case. Thus, from Eq. (5.40) we have

$$z'_M = (0.546 \times 60140 + 0.224 \times 290)/0.77 = 42{,}700$$

$$\frac{V_+^{1,2,\ldots,n}}{V_-^{1,2,\ldots,n}} = \frac{\ln\left(\dfrac{z'^B_W - z'_M}{z'^B_F - z'_M}\right)}{\ln\left(\dfrac{z'^A_W - z'_M}{z'^A_F - z'_M}\right)} = \frac{\ln\left[\dfrac{464 - 42{,}700}{290 - 42{,}700}\right]}{\ln\left[\dfrac{59{,}570 - 42{,}700}{60{,}140 - 42{,}700}\right]} = \frac{-0.00411}{-0.0332} = 0.124$$

	Feed mole fraction	Radioactive counts feed, z'_0	Radioactive counts product, z'_F
Isobutane	0.546	60140	59570
Isobutene	0.224	290	464
Hydrogen	0.230		
	1,000		

This experiment was conducted at a temperature of 430°C, and we wish to determine whether the rate-controlling steps in the process occur in the path of carbon transfer. Since $V_+/V_- < 1$, hydrogenation of the isobutene is taking place. If hydrogen is being adding or removed as atomically adsorbed species in a series of steps, it is reasonable to suppose that the stoichiometric number of each such step is unity. Then, if all the free energy change occurs in this series of steps, we have from Eq. (5.15):

$$V_+^{1,2,\ldots,n}/V_-^{1,2,\ldots,n} = v_{+1}v_{+2}\cdots v_n/v_{-1}v_{-2}\cdots v_n = \exp(-\Delta G/RT)$$

In order to test the hypothesis that this is the case, we must calculate the free energy change for the reaction

$$\exp(-\Delta G/RT) = K_p p_{iC_4H_{10}}/p_{H_2}p_{iC_4H_8}$$

The equilibrium constant, as determined experimentally, differs substantially from that reported in NBS thermodynamic tables. At 430°C the value obtained by Kamholz was 0.0117 atm. The average partial pressures for the reacting gases was

	Partial pressure (atm)
Isobutane	0.556
Isobutene	0.225
Hydrogen	0.231

Therefore $\exp(-\Delta G/RT) = (0.0117)(0.556)/(0.225)(0.231) = 0.125$. This is almost in exact agreement with $V_+^{1,2,\ldots,n}/V_-^{1,2,\ldots,n} = 0.124$, which is determined from the tracer experiment.

Some 30 runs were made by Kamholz with $V_+^{1,2,\ldots,n}/V_-^{1,2,\ldots,n}$ varying from 0.07 to 4.5. The low velocity ratios correspond to hydrogenation of isobutene whereas the high velocity ratios correspond to isobutane dehydrogenation. The values of the ratio

$$\frac{\exp(\Delta G/RT)}{(V_+^{1,2,\ldots,n}/V_-^{1,2,\ldots,n})} = 1.09 \pm 0.10$$

for this large range of values. This information can be interpreted as follows: assume a fairly general mechanistic model for the overall reaction $iC_4H_{10} = iC_4H_8 + H_2$:

Step	Elementary reaction	Stoichiometric number
1	$i\text{-}C_4H_{10} + l \underset{v_{-1}}{\overset{v_{+1}}{\rightleftharpoons}} i\text{-}C_4H_{10}l$	1
2	$i\text{-}C_4H_{10}l + l \underset{v_{-2}}{\overset{v_{+2}}{\rightleftharpoons}} i\text{-}C_4H_9l + Hl$	1
3	$i\text{-}C_4H_9l + l \underset{v_{-3}}{\overset{v_{+3}}{\rightleftharpoons}} i\text{-}C_4H_8l + Hl$	1
4	$i\text{-}C_4H_8l \underset{v_{-4}}{\overset{v_{+4}}{\rightleftharpoons}} i\text{-}C_4H_8 + l$	1
5	$2Hl \underset{v_{-5}}{\overset{v_{+5}}{\rightleftharpoons}} H_2l + l$	1
6	$H_2l \underset{v_{-6}}{\overset{v_{+6}}{\rightleftharpoons}} H_2 + l$	1

In that case we have

$$\exp\frac{-\Delta G}{RT} = \left(\frac{V_+^{1,2,3,4}}{V_-^{1,2,3,4}}\right)\left(\frac{V_+^{5,6}}{V_-^{5,6}}\right)$$

where $V_+^{1,2,3,4}/V_-^{1,2,3,4}$ corresponds to the determination of $V_+^{1,2,\ldots,n}/V_-^{1,2,\ldots,n}$ for carbon tracing so the ratio is equivalent to determining the value of

$$V_+^{5,6}/V_-^{5,6} = 1.09 \pm 0.10$$

If hydrogen adsorbs dissociatively, then steps 5 and 6 would be compressed into a single step. In either case the conclusions reached are that adsorption and desorption of hydrogen are very close to equilibrium, and the entire free energy decrease is sustained by transfer in the carbon path. If, for example, any large resistance were encountered in steps 5 and 6, the value of $V_+^{5,6}/V_-^{5,6}$ in the case of dehydrogenation would become larger than unity, and the ratio found for $\exp(-\Delta G/RT)/V_+^{1,2,\ldots,n}/V_-^{1,2,\ldots,n}$ by carbon tracing would greatly exceed unity.

It is generally supposed that hydrogen chemisorption is rapid at these temperatures, so perhaps the results obtained are not surprising. It is also reasonable to suppose that the stoichiometric number of each step involving carbon should be unity. This can, of course, be tested by assuming that $V_+^{5,6}/V_-^{5,6} = 1$ and calculating the apparent stoichiometric number from the expression

$$v_{app} = -\Delta G/RT/\ln(V_+^{1,2,\ldots,n}/V_-^{1,2,\ldots,n})$$

Using the same data as before we find $v_{app} = 1.02 \pm 0.07$.

This remarkable agreement may, in a sense, be considered as a test of the validity of transition-state theory as a means of relating thermodynamic to kinetic parameters.

The fact that the apparent stoichiometric number is close to one does not necessarily prove that there is only a single step that operates with a stoichiometric number of one. Any or all steps in the path of carbon transfer could be slow enough to exert an influence on the reaction rate, provided that the stoichiometric number of these steps were all unity.

Another technique employed in steady-state tracing of single-path reactions consists in operating a *closed* gradientless recirculating reactor at constant *volume* and observing rates of tracer transfer as functions of time. With time as the independent variable the system is said to be operated as a quasi-steady state, because it is generally the case that changes in tracer concentration of adsorbed species can be neglected. The equations of correlating data obtained by this technique involve time as the independent variable instead of distance through the catalyst bed. However, it is possible to observe concentrations at all times; whereas, in the case of plug-flow systems only terminal conditions can be obtained. The advantage of this mode of operation is that data are rapidly obtained for a wide range of conversions in a single experiment. The disadvantage is that it is necessary to operate at differential conversions and tracer transfer.

As indicated in the previous section, Eq. (5.2) must be replaced by

$$a_e^i = \frac{m^i \beta}{W} \frac{d(C^i z^i)}{dt} \tag{5.41}$$

Instead of Eq. (5.19) for the reaction $A \rightleftharpoons B$ we obtain

$$-\frac{\beta}{W}\frac{d(C^A z^A)}{dt} = \frac{\beta}{W}\frac{d(C^B z^B)}{dt} = V_+^{1,2} z^A - V_-^{1,2} z^B \tag{5.42}$$

The overall velocity V must still be taken as constant. Analogous to Eq. (5.23), we have

$$C^A = C_0^A - \frac{WV}{\beta}t \tag{5.43}$$

which corresponds to

$$V_+^{1,2} - V_-^{1,2} = V = -\frac{\beta}{W}\frac{dC^A}{dt} \tag{5.44}$$

By combining Eqs. (5.4.27) and (5.4.29), we obtain

$$\frac{V_+^{1,2}}{V_-^{1,2}} = 1 + (z^A - z^B)\frac{1}{C^A}\frac{dC^A}{dz^A} \tag{5.45}$$

Similar equations can be developed for cases in which more than one atom is marked in a species or more than one step occurs in a single path of transfer.

Whether we employ gradientless recirculating reactor systems, plug-flow reactors, or constant-volume recirculating systems in making such calculations, it is important that measurements be made under conditions so that the quantities in question can be readily determined. This means that conditions cannot be chosen either very close to or very far from thermodynamic equilibrium. At a great distance from equilibrium the determination of V_+/V_- becomes difficult because the reaction rate in one direction becomes negligible. On the other hand, close to equilibrium, the overall rate is negligible and $-\Delta G$ approaches zero.

In pioneering studies of the ammonia synthesis reaction conducted by Horiuti (1973) and his co-workers, the reaction was studied close to equilibrium, although this is not necessary. Later, at Horiuti's suggestion, Tanaka (1965, 1966) and co-workers (Tanaka and Matsuyama, 1971) conducted a series of studies for ammonia synthesis and decomposition under conditions further removed from equilibrium. The following example illustrates the use of this procedure, which indicates that nitrogen chemisorption, rather than one of the subsequent steps involving nitrogen, may be rate controlling, as was earlier concluded by Horiuti.

Example 5.4.2

The mechanism of the ammonia synthesis reaction over a singly promoted iron catalyst was studied by Tanaka and Matsuyama (1971) using ^{15}N as a tracer. Expressions used to determine the ratio of forward and reverse velocities in a constant-volume recirculating system were similar to Eq. (5.45). The ammonia synthesis reaction is assumed to consist of three steps:

(1) $N_2 + 2l \rightleftharpoons 2Nl$
(2) $H_2 + 2l \rightleftharpoons 2Hl$
(3) $3Hl + Nl \rightleftharpoons NH_3 + 4l$

5.4 Single-Path Reactions

Previous studies indicated that hydrogen chemisorption is extremely fast so the problem is to determine whether the first or third step is rate determining.

In the presence of a single rate-determining step the general relation

$$v_r = (-\Delta G/RT)/\ln(V_+/V_-)$$

applies. For the ammonia synthesis reaction

$$-\Delta G/RT = \ln(p_{H_2}^3 p_{N_2} K_p / p_{NH_3}^2)$$

The net reaction velocity V may be written

$$V = V_+ - V_- = \tfrac{1}{2} da/dt = -dn/dt$$

where a and n are the amounts of gaseous ammonia and gaseous nitrogen, respectively, assuming that the amount of ammonia adsorbed in the system does not change during the experiment. The rate of ^{15}N transfer from ammonia to nitrogen may be expressed as

$$d(z^A a)/dt = 2(z^N V_+ - z^A V_-)$$

or

$$d(z^N n)/dt = z^A V_- - z^N V_+$$

where z^A and z^N are the ^{15}N atomic fraction of the ammonia $^{15}NH_3/(^{14}NH_3 + {}^{15}NH_3)$ and of nitrogen $(^{14}N^{15}N + 2\,^{15}N_2)/2(^{14}N_2 + {}^{14}N^{15}N + {}^{15}N_2)$, respectively. In the same manner as used to obtain Eq. (5.45), expressions for v_+/v_- in terms of either dp_A/dz^A or dp_A/dz^N can be derived assuming that the concentration of ammonia a is proportioned to p_A, its partial pressure, and that of nitrogen n is proportioned to p_N. For the expression in terms of dp_A/dz^N:

$$V_+/V_- = 1 + (z^A - z^N)(1/2p_N)(dp_A/dz^N)$$

Combining this with the expression for the stoichiometric number, we have

$$r = \frac{-\Delta G/RT}{\ln(V_+/V_-)} = \frac{\ln(p_H^3 p_N K_p / p_A^2)}{\ln[1 + (z^A - z^N)(1/2p_N)(dp_A/dz^N)]}$$

Experiments were conducted with both the synthesis and decomposition of ammonia. For a typical synthesis run (Run No. 6), the temperature was 360°C, and the pressure was 850 mm Hg. In this series the initial ammonia pressure was 2.7 mm and the final pressure was 7.0 mm. Ammonia and nitrogen compositions were determined by mass-spectrometric observation. The pressure for p_H was taken as $p_H = 3(p_T - p_A)/4$ and for p_N was taken as $p_N = (p_T - p_A)/4$, where p_T is the total pressure. This corresponds to an initial mixture of stoichiometric proportions. At a given value of p_A the value of z^A and its derivative dp_A/dz^A were determined from a p_A versus z^A curve plotted from data taken during the course of the run.

Thus, for this run, at an intermediate observed p_A equal to 5 mm Hg after 21 hr of operation the following data were obtained:

$$z^A = 0.422, \quad P_N = 211 \text{ mm Hg}$$
$$z^N = 0.006, \quad P_H = 634 \text{ mm Hg}$$
$$dp_A/dz^N = 910 \text{ mm Hg}$$

This gives

$$V_+/V_- = 1 + (0.422 - 0.006)[1/(2 \times 211)](910) = 1.897$$

To obtain accurate values for v_N it was necessary to determine the equilibrium constant for the ammonia synthesis reaction. Values obtained did not agree exactly with those reported in the literature. For the runs at 360°C the equilibrium ammonia pressure, approached from both sides at times of over 200 hr, was 7.1 ± 0.1 mm Hg, which corresponded to a value of the equilibrium constant $K_P = (9.47 \pm 0.27) \times 10^{-10}$ mm Hg^{-2}. For the observation, corresponding to $P_A = 5$ mm Hg

$$-\Delta G/RT = \ln[((9.47 \times 10^{-10}) \times 211 \times 634^3)/5^2] = 0.711$$

From this, we calculate the value of v_r

$$v_r = 0.711/\ln 1.897 = 1.11$$

A number of additional runs are given in the original paper for both the synthesis and decomposition of ammonia.

From these results we conclude that the apparent stoichiometric number is one. If the mechanism follows what is assumed, this points to nitrogen chemisorption and desorption as being rate determining for ammonia synthesis and decomposition, respectively.

The catalyst employed in the above study was provided by Dr. C. Bokhoven and was a portion of the same preparation, containing 0.88% Al_2O_3, as employed by Bokhoven, and his co-workers (1959). It was about five times as active as the one previously used by Horiuti and co-workers and perhaps is more typical of commercial ammonia catalysts.

Studies using closed circulating systems have been conducted on the water-gas shift reaction employing ^{13}CO, ^{18}O, and D-labelled species. Results over an iron oxide catalyst are summarized by Oki and Mezaki (1973) and over a platinum catalyst by Masuda and Miyahara (1974), using approaches similar to those previously discussed. More work will be needed to completely elucidate the mechanisms involved. Oxygen-18 transfer is especially complicated because three species contain oxygen so that a single unbranched path does not exist for marked oxygen species.

The mechanism of oxidation of sulfur dioxide over platinum catalyst was studied in a similar system by Kaneko and Odanaka (1965) using ^{35}S as a tracer. The apparent stoichiometric number was found to be close to two, indicating that for this catalyst, oxygen chemisorption is not a rate-limiting step. For cases in which isotope transfer is not confined to a single path, the relationships involved will be discussed in the following section.

5.5 Multiple-Path Reactions

Where more than one path for atomic transfer exists, double levels of tracing will often enable more information to be developed regarding step velocities. Consider the reaction

$$A = B + C \tag{5.46}$$

where it is assumed that a single tracer element is employed that contains two atoms of tracer element in A and one atom each in B and C. Assume that the mechanism

$$A + l \underset{v_{-1}}{\overset{v_{+1}}{\rightleftharpoons}} Al, \quad Al \underset{v_{-2}}{\overset{v_{+2}}{\rightleftharpoons}} B + C + l \tag{5.47}$$

In this case the number of terminal species $n = 3$, the number of intermediates $k = 1$ so there will be $k + n - 1 = 3$ material balances:

$$\begin{aligned} -a^A &= 2z^A v_{+1} - 2z^{Al} v_{-1} \\ a^B &= z^{Al} v_{+2} - z^B v_{-2} \\ a^C &= z^{Al} v_{+2} - z^C v_{-2} \end{aligned} \tag{5.48}$$

Upon elimination of z^{Al} from these equations we obtain

$$\begin{aligned} \frac{V_+^{1,2}}{V_-^{1,2}} z^A - z^B - \frac{a^B}{V_-^{1,2}} + \left(\frac{a^A/2 + a^B}{v_{+1}}\right) \frac{V_+^{1,2}}{V_-^{1,2}} &= 0 \\ \frac{V_+^{1,2}}{V_-^{1,2}} z^A - z^C - \frac{a^C}{V_-^{1,2}} + \left(\frac{a^A/2 + a^C}{v_{+1}}\right) \frac{V_+^{1,2}}{V_-^{1,2}} &= 0 \end{aligned} \tag{5.49}$$

where $V_+^{1,2} = v_{+1} v_{+2}/(v_{-1} + v_{+2})$ and $V_-^{1,2} = v_{-1} v_{-2}/(v_{-1} + v_{+2})$.

Since

$$V = V_+^{1,2} - V_-^{1,2} = v_{+1} - v_{-1} = v_{+2} - v_{-2}$$

these equations contain two parameters, which could be specified as $V_+^{1,2}$ and v_{+1}. We could eliminate one of these parameters, say v_{+1}, from Eq. (5.49), leaving a single equation with one parameter, say $V_+^{1,2}$. A single-level experiment will enable this parameter to be determined, and substitution of $V_+^{1,2}$ into either equation of Eq. (5.49) will then enable v_{+1} to be determined. In turn we can then determine v_{+2}, so in this case all four of the unknown velocities in Eq. (5.48) can be established.

More complicated cases can be developed starting with the same basic equations, in which the constant velocity V is either retained or eliminated. The following example illustrates the method for a system requiring two levels of tracing.

Example 5.5.1.

The oxidation of sulfur dioxide over a vanadium catalyst was studied by Happel *et al.* (1973) using a gradientless recirculating reactor, with use of oxygen (^{18}O) and sulfur (^{35}S) isotopes for the tracing, assuming the same mechanism as in Example 5.3.1. A commercial vanadium oxide–sulfuric acid catalyst was used (typical analysis $V_2O_5 = 9.1\%$, $K_2O = 10.1$ wt.%, supplied by American Cyanamid Co.). Sulfur (^{35}S) was determined by scintillation counting and ^{18}O by mass spectrometry.

For this system we would like to determine, if possible, the velocities $v_{+1,2,3,4}$. For the mechanism, employing oxygen as the tracer element, the number of terminal species containing oxygen is $n = 3$ and the number of intermediates is $k = 3$ so the number of independent material balances is $n + k - 1 = 5$. They may be written as

$$-a_0^{O_2} = 2(z_0^{O_2} v_{+1} - z_0^{Ol} v_{-1})/v_1$$

$$-a_0^{SO_2} = 4(z_0^{SO_2} v_{+2} - z_0^{SO_2 l} v_{-2})/v_2$$

$$-a_0^{O_2} = 2(z_0^{Ol} v_{+3} - z_0^{SO_3 l} v_{-3})/v_3$$

$$-a_0^{SO_2} = 4(z_0^{SO_2 l} v_{+3} - z_0^{SO_3 l} v_{-3})/v_3$$

$$a_0^{SO_3} = 6(z_0^{SO_3 l} v_{+4} - z_0^{SO_3 l} v_{-4})/v_4$$

The subscript O is used to designate oxygen as the tracer element. The intermediates are eliminated from these equations; they are $z_0^{Ol}, z_0^{SO_3 l}$, and $z_0^{SO_3 l}$. For the reaction written as $O_2 + 2SO_2 = 2SO_3$, $v_1 = 1$; $v_2 = v_3 = v_4 = 2$. Thus, the following form can be obtained for the resulting two simultaneous equations:

$$\frac{V_+^{1,3,4}}{V_-^{1,3,4}}\left(\frac{a_0^{O_2}}{2} + Vz_0^{O_2}\right) = \frac{a_0^{O_2}}{2} + Vz_0^{O_2} + \frac{2}{3}\left(\frac{v_{+4}}{v_{-4}} - 1\right)\left(\frac{a_0^{O_2}}{2} - \frac{a_0^{SO_2}}{4}\right)$$

$$\frac{V_+^{2,3,4}}{V_-^{2,3,4}}\left(\frac{a_0^{SO_2}}{4} + Vz_0^{SO_2}\right) = \frac{a_0^{SO_2}}{2} + Vz_0^{SO_3} + \frac{1}{3}\left(\frac{v_{+4}}{v_{-4}} - 1\right)\left(\frac{a_0^{SO_2}}{4} - \frac{a_0^{O_2}}{2}\right)$$

5.5 Multiple-Path Reactions

Here v_{+4}/v_{-4} can be eliminated from these equations to obtain

$$\frac{V_+^{1,3,4}}{V_-^{1,3,4}}\left(\frac{a_0^{O_2}}{2} + Vz_0^{O_2}\right) + \frac{2V_+^{2,3,4}}{V_-^{2,3,4}}\left(\frac{a_0^{SO_2}}{4} + Vz_0^{SO_2}\right) = \frac{a_0^{SO_2}}{2} + \frac{a_0^{O_2}}{2} + 3Vz_0^{SO_3}$$

The last equation has two overall path velocities ratios as unknowns, which together with v_{+4}/v_{-4} constitute the three parameters that can be determined for the system.

The two parameters in the last equation can be determined by a two-level expression using ^{18}O tracing. We will illustrate the procedure by two runs with oxygen tracing using roughly equal overall velocities.

Run number	23	24
Wt.% of catalyst (g)	0.751	0.751
Reaction pressure (atm)	1.055	1.056
p_{N_2} (atm)	1.015	1.016
p_{O_2} (atm)	0.0223	0.0216
$p_{SO_2} \times 10^2$ (atm)	0.579	0.456
p_{SO_3} (atm)	0.0121	0.0136
$V \times 10^3$ [g mol/(g catalyst) (hr)]	1.229	1.458
$a_0^{O_2} \times 10^4$	-2.77	-0.0270
$a_0^{SO_2} \times 10^4$	0.607	-7.35
$a_0^{SO_3} \times 10^4$	2.16	7.38
$z_0^{O_2} \times 10^2$	10.69	0.313
$z_0^{SO_2} \times 10^2$	1.246	3.91
$z_0^{SO_3} \times 10^2$	0.700	1.742
$-\Delta G/RT$	3.513	2.775

First compute the following coefficients corresponding to run number 23

$$(a_0^{O_2}/2) + Vz_0^{O_2} = -1.385 \times 10^{-4} + 1.229 \times 10^{-3} \\ \times 10.69 \times 10^{-2} = -0.0712 \times 10^{-4}$$

$$(a_0^{SO_2}/2) + Vz_0^{SO_2} = 0.1518 \times 10^{-4} + 1.229 \times 10^{-3} \\ \times 1.246 \times 10^{-2} = 0.305 \times 10^{-4}$$

$$(a_0^{SO_2}/2) + (a_0^{O_2}/2) + 3Vz_0^{SO_3} = [(0.607/2 - 2.77/2)]10^{-4} \\ + (3 \times 1.229 \times 0.700)10^{-5} \\ = -0.823 \times 10^{-4}$$

Upon substitution of these values into the last equation, we obtain

$$-(V_+/V_-)^{1,3,4} + 8.56(V_+/V_-)^{2,3,4} = -11.56$$

A similar computation of the data for run number 24 gives

$$-(V_+/V_-)^{1,2,3} + 78.9(V_+/V_-)^{2,3,4} = 91.2$$

Solving these two equations, we find

$$(V_+/V_-)^{1,3,4} = 1.46\,(V_+/V_-)^{2,3,4} = 24.1$$

We can then calculate $v_{+4}/v_{-4} = -1.41$ from either of the original two equations obtained after elimination of concentrations of intermediates. The accuracy of the result would be increased by performing experiments at additional marking levels and correlating the data by standard statistical methods.

It is possible to avoid the use of double-oxygen-level tracing by use of ^{35}S marked species. The velocity ratio $(V_+/V_-)^{2,3,4}$ in the sulfur path can be determined directly based on a simple unbranched path. With $(V_+/V_-)^{2,3,4}$ determined from ^{35}S data, it is then possible to solve directly for $(V_+/V_-)^{1,3,4}$ and v_{+4}/v_{-4}.

One additional relationship among the step velocities is necessary to determine the velocities of all four steps. This is Eq. (5.15) derived from basic assumptions of thermodynamics and transition-state theory. In the present case, this is expressed as

$$\exp -\frac{\Delta G}{RT} = \frac{v_{+1}}{v_{-1}}\left(\frac{v_{+2}v_{+3}v_{+4}}{v_{-2}v_{-3}v_{-4}}\right) = \frac{K_p p_{SO_2}^2 p_{O_2}}{p_{SO_3}^2}$$

as developed in Example 5.3.1. If the velocity ratio $(V_+/V_-)^{2,3}$ is taken as a parameter, it is not necessary to use Eq. (5.15) as required for complete evaluation of all four step velocities.

Under all conditions, the data obtained indicate that step 1, oxygen chemisorption, is slowest. Further details are given in the original paper.

Following the studies of reactions in the isobutane–isobutene–hydrogen system using ^{14}C tracing described in Example 5.4.1, additional studies were made on the same system using deuterium tracing (Happel et al., 1973, 1976). It appears that slow steps in the mechanism correspond to $(V_+/V_-)^{1,2}$, involving isobutane chemisorption to form iC_4H_9l. This conclusion is uncertain because of problems in data analysis, including the fact that two types of $i\text{-}C_4H_9l$ are possible. With n-butane dehydrogenation the situation would be still more complicated.

Considerable information about catalytic mechanisms can be obtained by steady-state tracing, especially when additional use is made of the transition state theory relationship and when velocity groupings are employed to reduce the number of required parameters. With branching mechanisms the picture is more complete than for a series of steps in an unbranched path.

However, the transient superposition technique discussed in Chapter 7 is a more powerful tool for the study of reaction mechanisms. It not only enables assessment to be made of the velocities of reaction steps, but estimates of the surface concentrations of adsorbed species, if they are present in sufficient concentrations.

List of Symbols

a_e^i	Transfer rate of traced element e to species i, (e may be omitted when no ambiguity exists) (g_{atom}/g_{cat})
C^i	Concentration of species i (fraction)
F^i	Rate of removal of species i at any point from a recirculating reactor (moles/sec)
F_F^i	Rate of flow of species i to a reactor (moles/sec)
F_w^i	Rate of removal of species i from a plug-flow reactor (moles/sec)
ΔG	Gibbs free-energy change for overall reaction
Δg	Gibbs free-energy change for mechanistic step
K_p	Equilibrium constant in pressure units
m^i	Number of atoms of tracer element in molecule of i
R	Gas constant
T	Absolute temperature (K)
V	Net overall reaction velocity per unit mass of catalyst
$V_+^{1,2,3,\ldots,n}$	Overall forward velocity per unit mass
$V_-^{1,2,3,\ldots,n}$	Overall reverse velocity per unit mass
W	Weight of catalyst (g)
w	Variable weight traversed in plug-flow reactor
z_e^i	Fraction of tracer element e in effluent of recirculating reactor or at any point in plug-flow reactor (e may be omitted when no ambiguity exists)
z_{Fe}^i	Fraction of tracer element e in feed to reactor (e may be omitted when no ambiguity exists)
z_{We}^i	Fraction of tracer element e leaving plug-flow reactor (e may be omitted when no ambiguity exists)
β	Void volume in reactor (ml)
v_i	Stoichiometric number of ith step

References

Bokkoven, C., Gorgels, M. J., and Mars, P. (1959), *Trans. Faraday Soc.* **55**, 315.
Emmett, P. H. (1972). *Catal. Rev.* **7**, 1.
Eyring, H. J. (1935). *Chem. Phys.* **3**, 107.
Happel, J. (1972). *Catal. Rev.* **6**, 221.
Happel, J. (1980). *J. Res. Inst. Catal. Hokkaido Univ.* **28**, 185.
Happel, J., and Hnatow, M. A. (1976). *J. Catal.* **42**, 54.
Happel, J., Hnatow, M. A., and Rodriguez, A. (1973). *AIChE J.* **19**, 1075.

Happel, J., Kamholz, K., Walsh, D., and Strangio, V. (1973a). *Ind. Eng. Chem. Fundam.* **12**, 263.
Happel, J., Hnatow, M. A., Strangio, V. A. (1976). *Ind. Eng. Chem. Fundam.* **15**, 115.
Hollingsworth, C. A. (1957). *J. Chem. Phys.* **27**, 1346.
Horiuti, J. (1973). *Ann. N. Y. Acad. Sci.* **213**, 5.
Horiuti, J., and Ikusima, M. (1939). *Proc. Imp. Acad. (Tokyo)* **15**, 39.
Horiuti, J., and Nakamura, T. (1967). *Adv. Catal.* **17**, 1.
Kaneko, Y., and Odanaka, H. (1965). *J. Res. Inst. Catal. Hokkaido Univ.* **13**, 29.
Kamholz, K. (1970). Doctoral thesis, School of Engineering and Science, New York University, New York.
Masuda, M., and Miyahara, K. (1974). *Bull. Chem. Soc. Jpn.* **47** (5), 1058.
Oki, S., and Mezaki, R. (1973). *J. Phys. Chem.* **77**, 1601.
Park, S. W., and Himmelblau, D. M. (1982). *Chem. Eng. J.* **25**, 163.
Temkin, M. I. (1979). *Adv. Catal.* **28**, 173.
Tanaka, K. (1965). *J. Res. Inst. Catal. Hokkaido Univ.* **13**, 119.
Tanaka, K. (1966). *J. Res. Inst. Catal. Hokkaido Univ.* **14**, 153.
Tanaka, K., and Matsuyama, A. (1971). *J. Res. Inst. Catal. Hokkaido Univ.* **19**, 63.

Chapter 6

Identifiability and Distinguishability

6.1 Introduction

Ideally, before modeling transient tracing data, it would be desirable to determine from the structure of each model proposed whether in principle the desired parameters could be determined uniquely and distinguished from those of competing models. The methods for considering these problems are presented in this chapter.

However, transient behavior is more complex than the case for steady-state modeling so that it might be possible to expend considerable effort in resolving identifiability of structures that turn out, in fact, to be far from acceptable in actual data correlation. Thus, when a model does not conform to structures already previously developed, it may be more practical to proceed directly to computer correlation, as discussed in detail in Chapter 7.

As discussed in Sections 4.4 and 4.5, when an appropriate model has been assumed, it is possible to set up the differential equations with desired parameters (surface concentrations and step velocities) as constants.

The procedure for determining parameters is to write material balances to describe the transient concentrations of marked species. Such models can often be expressed in terms of linear differential equations with constant coefficients as given by Eq. (4.7). The following is a more generalized version of the systems equations:

$$\mathcal{M}_0: dz/dt = A(\theta)z + Bu \tag{6.1}$$

$$y = Cz \tag{6.2}$$

where **z** is the vector of state variables corresponding to the fraction of tracer in a given atomic species; t, the time following initial injection of traced species to a system at steady state with respect to the chemical reaction taking place; θ, the vector of the parameters to be estimated, which are the velocities of mechanistic steps and the unknown concentrations of surface intermediates; **u**, the vector of input of tracer to the system (in studies of chemical mechanisms u is usually a scalar since there is only single input to either a species entering into the reactor or a reaction product); **y**, the vector of the observed output tracer fraction (this will usually consist of the values corresponding to terminal species; if it is possible to observe transients of intermediates, this vector could be expanded accordingly); **A**, a matrix consisting of constants appearing in the material balance for each compartment; **B**, a matrix determined from tracer input (in chemical mechanism studies this will usually be a vector involving a single nonzero constant); and **C**, a matrix of constants corresponding to observed terminal species.

If the differential equations [(6.1) and (6.2)] can be solved analytically, it is possible to determine the groupings containing the parameters where θ is defined by

$$\theta = [C_i, v_{\pm j}]T \tag{6.3}$$

where $\pm j$ represents forward and back reactions of corresponding steps.

However, even for very simple heterogeneous reactions analytical solution of the equations becomes difficult. It is desirable to consider the structure of the systems, as discussed in the following sections.

6.2 Identifiability Using Transient Tracing

The problem of structural identifiability has been discussed by many authors. For a detailed bibliography see Walter (1982). Most of the papers originate from econometrics or the biosciences. The notion has been discussed in a chemical context by Park and Himmelblau (1982). These authors described criteria for models represented by algebraic equations such as the Langmuir–Hinshelwood–Hougen–Watson (LHHW) rate equations as well as models represented by differential equations. A model will be structurally globally identifiable (sgi) (Lecourtier and Walter, 1981) if the identities of the "model" $\mathcal{M}(\theta)$ and of the "process" $\mathcal{M}(\theta)$ imply that the parameters $\hat{\theta}$ of the identified model are equal to the "true" parameters of the process for almost any θ belonging to the admissable parameter set Θ. In this context the term process simply refers to a proposed structure and does not imply the correctness of the structure from a chemical viewpoint. Since the structure of these models does not depend on any actual data, it is only necessary to

6.2 Identifiability Using Transient Tracing

assume that, if a given model is really viable, a set of parameters does exist. A model $\mathcal{M}(\theta)$ is structurally globally identifiable (sgi) if and only if, for all values of its parameter vector θ (except possibly for some very special values), there exists an input function u such that the relationship

$$\mathbf{y} = \mathbf{y}(t, \theta, \mathbf{u}) = \mathbf{y}(t, \hat{\theta}, \mathbf{u}) \quad \text{for all} \quad t \geq 0 \quad (6.4)$$

implies that $\theta = \hat{\theta}$. Structural global identifiability is a property of the model itself that is determined independently of any tracer measurements.

Local structural identifiability is a weaker condition than global structural identifiability. A model $\mathcal{M}(\theta)$ is structurally locally identifiable (sli) if and only if, under the same assumptions, Eq. (6.4) implies that θ belongs to a denumerable set. For practical application this set should be finite and small.

For any reasonable input (a step function, for example) there is only one unique parameter vector that corresponds to a given output behavior for an sgi model. On the other hand, for an sli model, a number of parameter vectors exist corresponding to exactly the same output behavior.

Walter and Lecourtier (1982) present several methods that are suitable for testing identifiability in state-space models including linear time-invariant models described by Eqs. (6.1) and (6.2). One of these, termed the transfer function approach (see Happel et al., 1984; Walter et al. 1985) is especially applicable for models with a small number of inputs and outputs and a highly constrained structure with many nonexistent exchanges between compartments. This is the situation encountered in the modeling of catalytic systems by transient tracing of terminal species. The method is similar to that originally proposed by Bellman and Åström (1970) and takes advantage of the fact that the transfer function employed in control theory yields the parameter groupings that appear in the solutions of Eqs. (6.1) and (6.2) without actually requiring their explicit solution. This method is detailed in this chapter, although the others may also be useful in some situations.

The transfer function is obtained by taking the Laplace transforms of the systems equations. Then in Laplace transform space the response becomes

$$\bar{\mathbf{y}}(s) = \mathbf{H}(s, \theta)\bar{\mathbf{u}}(s) \quad (6.5)$$

where $\mathbf{H}(s, \theta)$ is the transfer function, which is defined as

$$\mathbf{H}(s, \theta) = \mathbf{C}(s\mathbf{I} - \mathbf{A})^{-1}\mathbf{B} \quad (6.6)$$

where $\mathbf{I}$ is the identity matrix. The elements of $\mathbf{H}(s, \theta)$ are ratios of the polynomials in s. The unknown parameter vector θ is sgi, if for almost any value of θ belonging to Θ,

$$\mathbf{C}(\theta)[s\mathbf{I} - \mathbf{A}(\theta)]^{-1}\mathbf{B}(\theta) = \mathbf{C}(\hat{\theta})[s\mathbf{I} - \mathbf{A}(\hat{\theta})]^{-1}\mathbf{B}(\hat{\theta}) \quad \text{for all} \quad s \quad (6.7)$$

implies that $\theta = \hat{\theta}$. That is, the vector θ is sgi if different values of θ do not give the same input–output behavior. Structural local identifiability implies a restricted parameter space such that

$$\|\theta - \tilde{\theta}\|\delta, \quad \delta > 0 \tag{6.8}$$

The actual implementation of the procedure for applying the transfer function approach can be best understood by following worked out examples. First we will consider a simple system for the reaction $A = B$ given in Section 4.5 with definition sketch, Fig. 4.1.

Example 6.2.1

Assume that the species A and B contain a single atom that can be traced. If a step-function input of traced A is introduced at time $t = 0$ and the marking level in A is observed, we would like to determine whether the system is sgi, assuming the following mechanism:

$$\begin{aligned} A + l &\underset{v_{-1}}{\overset{v_{+1}}{\rightleftharpoons}} Al \\ Al &\underset{v_{-2}}{\overset{v_{+2}}{\rightleftharpoons}} B + l \end{aligned} \tag{6.9}$$

where l represents catalyst sites for chemisorption and Al is a chemisorbed intermediate. The reaction velocities for the two mechanistic steps at steady state are given by $v_{\pm i}$ ($i = 1, 2$). It is assumed that for this example the feed is pure component A and that all the forward and reverse step velocities are nonzero (at steady state but not at equilibrium). Note that the compartments need not be physically distinct phases. Thus, the compartments A and B coexist in the gas phase.

Material balances following the notation of Happel (1974) are

$$\begin{aligned} (\beta C^A/W)\dot{z}^A &= -[F^A/W) + v_{+1}]z^A + v_{-1}z^{Al} + (F_F^A/W)u \\ C^{Al}\dot{z}^A &= v_{+1}z^A - (v_{-1} + v_{+2})z^{Al} + v_{-2}z^B \\ (\beta C^B/W)\dot{z}^B &= v_{+2}z^{Al} - [v_{-2} + (F^B/W)]z^B \end{aligned} \tag{6.10}$$

where C^{Al} is the concentration of Al on the catalyst surface; C^A, C^B, the concentration of A and B in the gas phase; F_F^A, the rate of feed of A to the system; F^A, the rate of removal of A from the system; F^B, the rate of removal of B from the system; β, the dead volume of the system; W, the weight of catalyst; $z_F^A = u$, the fractional marking of the feed; and z^A, z^{Al}, z^B, all the fractional markings of A, Al and B.

The assumption that the overall reaction is at steady state allows us to write the following equation:

$$(F_F^A/W) - (F^A/W) = v_{+1} - v_{-1} = v_{+2} - v_{-2} = F^B/W$$

6.2 Identifiability Using Transient Tracing

No term appears for F_F^B because B is not assumed to be present in the feed. Otherwise the equations repeat those in Chapter 4.

It is convenient to combine known constants separately, designating the parameters of interest as

$$\theta_1 = v_{-1}, \qquad \theta_2 = v_{-2}, \qquad \theta_3 = C^A$$

with

$$k_0 = F^B/W, \qquad k_1 = F^A/W, \qquad k_2 = \beta C^A/W, \qquad k_3 = \beta C^B/W$$

and we then obtain

$$v_{+1} = k_0 + \theta_1, \qquad v_{+2} = k_0 + \theta_2, \qquad F_F^A/W = k_0 + k_1$$

Equation (6.10) may be written as

$$\dot{z} = \begin{bmatrix} \dfrac{-(k_1+k_0+\theta_1)}{k_2} & \dfrac{\theta_1}{k_2} & 0 \\ \dfrac{k_0+\theta_1}{\theta_3} & -\dfrac{(\theta_1+k_0+\theta_2)}{\theta_3} & \dfrac{\theta_2}{\theta_3} \\ 0 & \dfrac{k_0+\theta_2}{k_3} & -\dfrac{(\theta_2+k_0)}{k_3} \end{bmatrix} z + \begin{bmatrix} \dfrac{k_0+k_1}{k_2} \\ 0 \\ 0 \end{bmatrix} u \qquad (6.11)$$

$$y = [1 \; 0 \; 0] z$$

It is possible to determine $\bar{y}(s)$ by taking the Laplace transforms of Eq. (6.11) and solving the three simultaneous linear algebraic equations for z^A. Using the relation of Eq. (6.10), the solution may be written as

$$\bar{y}(s) = \mathbf{C} \cdot (s\mathbf{I} - \mathbf{A})^{-1} \cdot \mathbf{B} \cdot \bar{u}(s) \qquad (6.12)$$

In view of the particular form of the **B** and **C** matrices, Eq. (6.12) becomes

$$\bar{y}(s) = [(k_0 + k_1)/k_2] a_{11}^{(-1)}(s) \bar{u}(s) \qquad (6.13)$$

where $a_{11}^{(-1)}(s)$ is the entry in the first row and the first column of the matrix $(s\mathbf{I}-\mathbf{A})^{-1}$. Thus, the transfer function may be easily deduced from Eq. (6.11)

$$\mathbf{H}(s, \theta) = \frac{k_0 + k_1}{k_2} a_{11}^{(-1)}(s) = \frac{N(s, \theta)}{D(s, \theta)} \qquad (6.14)$$

with

$$N(s, \theta) = \frac{k_0 + k_1}{k_2} \left[s^2 + \left(\frac{\theta_1 + \theta_2 + k_0}{\theta_3} + \frac{\theta_2 + k_0}{k_3} \right) s + \frac{(\theta_1 + k_0)(\theta_2 + k_0)}{\theta_3 k_3} \right]$$

$$(6.15)$$

and

$$D(s, \theta) = s^3 + \left(\frac{k_0 + k_1 + \theta_1}{k_2} + \frac{\theta_1 + \theta_2 + k_0}{\theta_3} + \frac{\theta_2 + k_0}{k_3}\right) s^2$$

$$+ \left[\frac{(\theta_1 + k_0)(\theta_2 + k_0)}{\theta_3 k_3} + \frac{(\theta_1 + \theta_2 + k_0)(k_0 + k_1 + \theta_1) - \theta_1(k_0 + \theta_1)}{k_2 \theta_3}\right.$$

$$\left. + \frac{(\theta_2 + k_0)(k_1 + k_0 + \theta_1)}{k_2 k_3}\right] s + \frac{(\theta_1 + k_0)(\theta_2 + k_0)(k_1 + k_0)}{\theta_3 k_2 k_3} \quad (6.16)$$

Identifying $H(s, \theta)$ with $H(s, \hat{\theta})$ as in Eq. (6.16) we find that $\mathcal{M}(\theta)$ anf $\mathcal{M}(\hat{\theta})$ are output indistinguishable if and only if the following set of nonlinear algebraic equations has a solution $\theta = \hat{\theta}$. These equations are derived by extracting the coefficients of the various polynomials in s from Eqs. (6.2.13) and (6.2.14) assuming that k_i are known:

$$\frac{\theta_1 + \theta_2 + k_0}{\theta_3} = \frac{\theta_2}{k_3} = \frac{\hat{\theta}_1 + \hat{\theta}_2 + k_0}{\hat{\theta}_3} + \frac{\hat{\theta}_2}{k_3} \quad (6.17)$$

$$\frac{(\theta_1 + k_0)(\theta_2 + k_0)}{\theta_3} = \frac{(\hat{\theta}_1 + k_0)(\hat{\theta}_2 + k_0)}{\hat{\theta}_3} \quad (6.18)$$

$$\frac{\theta_1}{k_2} + \frac{\theta_1 + \theta_2 + k_0}{\theta_3} + \frac{\theta_2}{k_3} = \frac{\hat{\theta}_1}{k_2} + \frac{\hat{\theta}_1 + \hat{\theta}_2 + k_0}{\hat{\theta}_3} + \frac{\hat{\theta}_2}{k_3} \quad (6.19)$$

$$\frac{(\theta_2 + k_0)(k_0 + k_1 + \theta_1) + \theta_1 k_1}{\theta_3} + \frac{(\theta_2 + k_0)(k_1 + k_0 + \theta_1)}{k_3}$$

$$= \frac{(\hat{\theta}_2 + k_0)(k_0 + k_1 + \hat{\theta}_1) + \hat{\theta}_1 k_1}{\hat{\theta}_3} + \frac{(\hat{\theta}_2 + k_0)(k_1 + k_0 + \hat{\theta})}{k_3} \quad (6.20)$$

Note that the constant term of $D(s, \theta)$ in Eq. (6.16) leads to the same equation as the constant term of $N(s, \theta)$, i.e., Eq. (6.18). Also Eq. (6.20) follows because of the term of degree one in s in $D(s, \theta)$. In this equation the term $(\theta_1 + k_0)(\theta_2 + k_2)/\theta_3$ does not appear because it is already present in Eq. (6.18).

The test for structural identifiability is thus a problem of solving the four equations (6.17)–(6.20), which can easily be treated by hand here. In more complicated cases symbolic manipulation routines may be required (Walter and Lecourtier, 1982). Here, the difference between Eq. (6.19) and Eq. (6.17) leads to

$$\theta_1 = \hat{\theta}_1 \quad (6.21)$$

Then, Eqs. (6.21) and (6.18) yield

$$(\theta_2 + k_0)/\theta_3 = (\hat{\theta}_2 + k_0)/\hat{\theta}_3 \quad (6.22)$$

6.3 Distinguishability Using Transient Tracing

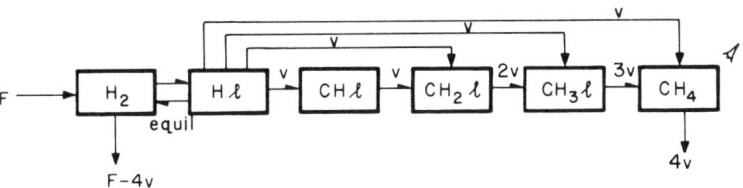

FIG. 6.1 Deuterium tracing of methanation model.

If these two relationships are substituted into Eq. (6.18), we obtain taking Eq. (6.17) into account,

$$\theta_2 = \hat{\theta}_2 \qquad (6.23)$$

and finally, by substitution of Eq. (6.23) into (6.22) we obtain

$$\theta_3 = \hat{\theta}_3 \qquad (6.24)$$

Thus, this model is sgi. That is, if a reaction can be accurately modeled by Eq. (6.9) the parameters might, with noise-free data, be uniquely determined. This contrasts with Example 4.5.1 in which the model could not be identified by observing compartment A alone, because the reaction steps were taken to be irreversible in that case.

Models of this type are of general interest because they apply not only to isomerization reactions but to any system that involves a path of transfer of marked species without addition or removal of tracers from intermediate compartments in a chain. They are often described as *catenary*. It can be shown that for a catenary model with any number of compartments, if the initial compartment or *head* is injected and observed the model will be sgi provided that all steps are reversible. For the conditions assumed, with known rates of feed and product and known concentrations in the gas phase, a model with only the tail (i.e., last compartment) observed is not identifiable if the model has more than three compartments (Happel et al., 1986).

Different situations arise when the tracer species added moves stepwise to a reacting molecule in a series of steps. Such a system is shown in Fig. 6.1. In this system deuterium substitution is employed to study the hydrogenation of carbon monoxide over a nickel catalyst to produce methane. Details of the actual experiments are given by Happel et al. (1982). This model can be shown to be sgi, and, in fact, values have been reported for the parameters consisting of surface concentrations of adsorbed intermediates (see Example 7.4.1).

6.3 Distinguishability Using Transient Tracing

Given a set of identifiable models, discrimination often involves experimental design to select the model that best fits the data. Methods for accomplishing this have been discussed (see e.g., Beck and Arnold, 1977). It often happens that discrimination among a large number of proposed models is not clear. One is confronted by the problem that even if it is possible to identify the parameters precisely for each model, it may also be possible from the structure of the models themselves that more than one model may fit a given set of data equally well. Then even careful experimentation would not serve as a means for deciding which model is best.

An approach to this problem is to examine the structure of the proposed mechanisms using principles similar to those discussed in the previous section for studying identifiability. Thus, we may seek to determine whether two given models can be distinguished from each other. The procedure is then repeated for all possible pairs of an assumed set to determine the extent to which they can be distinguished. The basic procedure is set forth by Walter and Lecourtier (1982) and Walter et al. (1984a,b, 1986) as follows.

Assume that there exists a set of parameters for a model $\mathcal{M}(\theta)$ and that we wish to examine the possibility that a new model $\hat{\mathcal{M}}(\hat{\theta})$ could be satisfied by the same set of parameters. The new model $\hat{\mathcal{M}}$ will be structurally output distinguishable from $\mathcal{M}$ if and only if for all values of the parameter vector θ (except possibly for some very special values) there exists an input function $\mathbf{u}$ such that the equation

$$\hat{y}(\mathbf{t}, \hat{\theta}, \mathbf{u}) = y(t, \theta, \mathbf{u}) \qquad \text{for all} \quad t \geq 0 \tag{6.25}$$

which implies that $\hat{\theta} \neq \theta$.

The problem here is more complicated than in the case of identifiability studies. It is necessary first to test for the distinguishability of $\hat{\mathcal{M}}$ from $\mathcal{M}$ as previously discussed. In actual modeling, however, we could just as well have arrived at model $\hat{\mathcal{M}}$ first. So it would then be necessary to test the distinguishability of $\mathcal{M}$ from $\hat{\mathcal{M}}$. Thus, for complete structural distinguishability it is necessary to test both the distinguishability of $\hat{\mathcal{M}}$ from $\mathcal{M}$ and then the distinguishability of $\mathcal{M}$ from $\hat{\mathcal{M}}$. If both tests show indistinguishability the structures of the models are said to be structurally distinguishable (sd). If such is the case the two models cannot have the same output behavior.

As in the case of Eq. (6.6), the use of the transfer function approach allows elimination of time and inputs. As shown by Walter et al. (1984a,b) structural identifiability of the models is neither necessary or sufficient for them to be sd. Thus, it may be possible to determine the most likely model without being able to determine all the parameters comprising the model. We will illustrate the method by considering the distinguishability of two models similar to that in Example 6.2.1.

Example 6.3.1

For a reaction similar to that considered in Example 6.2.1, we wish to determine whether two models are distinguishable. In one model, the first step

6.3 Distinguishability Using Transient Tracing

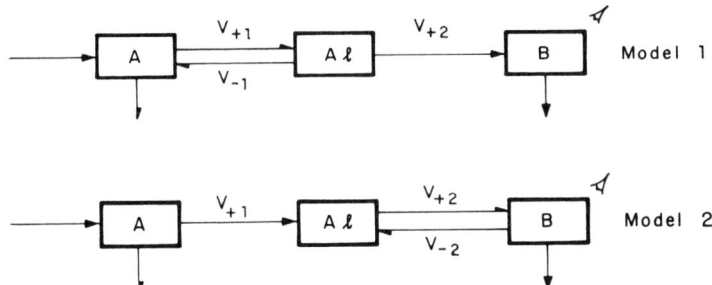

FIG. 6.2 Simple isomerization model, Example 6.3.1: distinguishability.

will be taken as unidirectional, and in the other model the second step will be taken as unidirectional instead of assuming both steps to be reversible, as was the case for Example 6.2.1. The definition sketch is Fig. 6.2. In both cases the head compartment A is injected with tracer but in this example the tail compartment B is observed.

In order to test the distinguishability of the two models, it is necessary to calculate the transfer function for each of them in the same manner as in Example 6.2.1. For Model 1, we take $v_{-2} = 0 = \theta_2$ and

$$\dot{z} = \begin{bmatrix} \dfrac{-(k_1 + k_0 + \theta_1)}{k_2} & \dfrac{\theta_1}{k_2} & 0 \\ \dfrac{k_0 + \theta_1}{\theta_3} & \dfrac{-(\theta_1 + k_0)}{\theta_3} & 0 \\ 0 & \dfrac{k_0}{k_3} & \dfrac{-k_0}{k_3} \end{bmatrix} z + \begin{bmatrix} \dfrac{k_0 + k_1}{k_2} \\ 0 \\ 0 \end{bmatrix} u \quad (6.26)$$

$$\mathbf{y} = [0\ 0\ 1]\mathbf{z}$$

Note that because of the observation of the B species $\mathbf{y} = [1\ 0\ 0]\mathbf{z}$ in Eq. (6.11) is replaced by $\mathbf{y} = [0\ 0\ 1]\mathbf{z}$ in Eq. (6.26) and then Eq. (6.13) becomes

$$\bar{y}(s) = [(k_0 + k_1)/k_2] a_{31}^{(-1)}(s) \quad (6.27)$$

The transfer function associated with the model $\mathcal{M}$ may be written as

$$\mathbf{H}(s, \boldsymbol{\theta}) = N_1(s, \boldsymbol{\theta})/D_1(s, \boldsymbol{\theta}) \quad (6.28)$$

with $N_1(s, \boldsymbol{\theta}) = [(k_0 + k_1)/k_2](k_0/k_3)[(k_0 + \theta_1)/\theta_3]$ and $D_1(s, \boldsymbol{\theta}) = D(s, \boldsymbol{\theta})|_{\theta_2 = 0}$, where $D(s, \boldsymbol{\theta})$ is defined by Eq. (6.16).

Similarly for the transfer function associated with the second model $\hat{\mathcal{M}}$ we take $v_{-1} = 0 = \hat{\theta}_1$. Thus we have

$$\bar{\mathbf{H}}(s, \hat{\boldsymbol{\theta}}) = N_2(s, \hat{\boldsymbol{\theta}})/D_2(s, \hat{\boldsymbol{\theta}}) \quad (6.29)$$

with

$$N_2(s,\hat{\theta}) = \frac{k_0 + k_1}{k_2} \frac{k_0}{\hat{\theta}_3} \frac{k_3 + \hat{\theta}_2}{k_3}$$

Note that H depends on θ_1 and θ_3 and $\hat{H}$ depends on $\hat{\theta}_2$ and $\hat{\theta}_3$.

In order to test the distinguishability of $\mathcal{M}(\theta)$ and $\hat{\mathcal{M}}(\theta)$, we must write the equality of the two transfer functions $H(s,\theta) = \hat{H}(s,\hat{\theta})$ for all s, which is equivalent to the following set of equations:

$$N_1(s,\theta) = N_2(s,\hat{\theta}) \Rightarrow \frac{k_0 + \theta_1}{\theta_3} = \frac{k_3 + \hat{\theta}_2}{\hat{\theta}_3} \qquad (6.30)$$

$$D_1(s,\theta) = D_2(s,\hat{\theta}) \Rightarrow \begin{cases} \dfrac{\theta_1}{k_2} + \dfrac{\theta_1 + k_0}{\theta_3} = \dfrac{\hat{\theta}_2 + k_0}{\hat{\theta}_3} + \dfrac{\hat{\theta}_2}{k_3} & (6.31) \\[2mm] \dfrac{(\theta_1 + k_0)k_0}{\theta_3 k_3} + \dfrac{(\theta_1 + k_0)(k_0 + k_1)}{k_2 \theta_3} + \dfrac{k_0(k_1 + k_0 + \theta_1)}{k_2 k_3} \\[2mm] \quad = \dfrac{k_0(\hat{\theta}_2 + k_0)}{\hat{\theta}_3 k_3} + \dfrac{(\hat{\theta}_2 + k_0)(k_0 + k_1)}{k_2 \hat{\theta}_3} \\[2mm] \quad + \dfrac{(\hat{\theta}_2 + k_0)(k_1 + k_0)}{k_2 k_3} & (6.32) \end{cases}$$

Note that the equality of constant terms of $D_1(s,\theta)$ and $D_2(s,\hat{\theta})$ repeats Equation (6.30).

In order to test whether $\mathcal{M}(\theta)$ is distinguishable from $\hat{\mathcal{M}}(\hat{\theta})$, we must look for solutions of Eqs. (6.30)–(6.32) with respect to θ_1 and θ_3, and in order to test whether $\hat{\mathcal{M}}(\hat{\theta})$ is distinguishable from $\mathcal{M}(\theta)$, we must look for the solutions of the same equations with respect to $\hat{\theta}_2$ and $\hat{\theta}_3$. Here it is easy to simplify the set of equations. By subtracting Eq. (6.30) from Eq. (6.31) we obtain

$$\theta_1/k_2 = \hat{\theta}_2/k_3 \qquad (6.33)$$

By subtracting the product of $(k_0/k_3) + (k_0 + k_1)/k_3$, times Eq. (6.30) from Eq. (6.32) and simplifying, we obtain

$$k_0\theta_1 = (k_0 + k_1)\hat{\theta}_2 \qquad (6.34)$$

Except for special values of $k_0, k_1, k_2,$ and k_3, (6.33) and (6.34) have no solution either with respect to θ_1 or with respect to $\hat{\theta}_2$. Therefore, $\mathcal{M}(\theta)$ and $\mathcal{M}(\hat{\theta})$ are distinguishable. Also in this case, model 1 is not sgi, whereas model 2 is. Thus, we see that the models can be sd without being sgi.

6.4 General Procedure for Model Testing

The procedures given in this chapter for the study of identifiability and distinguishability can be applied in a systematic way to determine to what extent proposed isotopic tracer experiments can lead to the unambiguous choice of a reaction mechanism for those considered possible on the basis of an enumeration such as that advanced by Happel and Sellers (1982, 1983).

To show how this can be accomplished, the following example follows that detailed for ammonia synthesis by Happel and Sellers (1983, Example 3).

Example 6.4.1

For the catalytic synthesis of ammonia, all possible mechanisms were developed for the reaction

$$N_2 + 3H_2 = 2NH_3$$

to be consistent with the following choice of elementary reaction steps considered as plausible candidates to participate in any mechanism:

$$
\begin{aligned}
s_1: & \quad N_2 + l \rightleftharpoons N_2 l \\
s_2: & \quad N_2 l + H_2 \rightleftharpoons N_2 H_2 l \\
s_3: & \quad N_2 H_2 l + l \rightleftharpoons 2NHl \\
s_4: & \quad N_2 + 2l \rightleftharpoons 2Nl \\
s_5: & \quad Nl + Hl \rightleftharpoons NHl + l \\
s_6: & \quad NHl + Hl \rightleftharpoons NH_2 l + l \\
s_7: & \quad NHl + H_2 \rightleftharpoons NH_3 + l \\
s_8: & \quad H_2 + 2l \rightleftharpoons 2Hl \\
s_9: & \quad NH_2 + Hl \rightleftharpoons NH_3 + 2l
\end{aligned}
\tag{6.35}
$$

As above, the symbol l in Eqs. (6.35) refers to active surface sites on the catalyst. Every species with an l in it is an intermediate and the rest are terminal species. It was shown that there are six possible mechanisms for the synthesis reaction, as listed in Table 6.1. Mechanisms m_1 and m_4 are proposed by Horiuti (1973) and Temkin (1979), who did extensive studies of ammonia synthesis. The steps chosen for Eq. (6.35) are a combination of all those proposed by these two scientists.

We wish to determine how tracing with ^{15}N could serve to discriminate among these six mechanisms. Figure 6.3 shows the path of atomic nitrogen transfer for the six mechanisms. It is assumed that traced N_2 species is injected into the system in all cases and that the rate of production of ammonia as well as terminal concentrations of species are known. The transient of ^{15}N

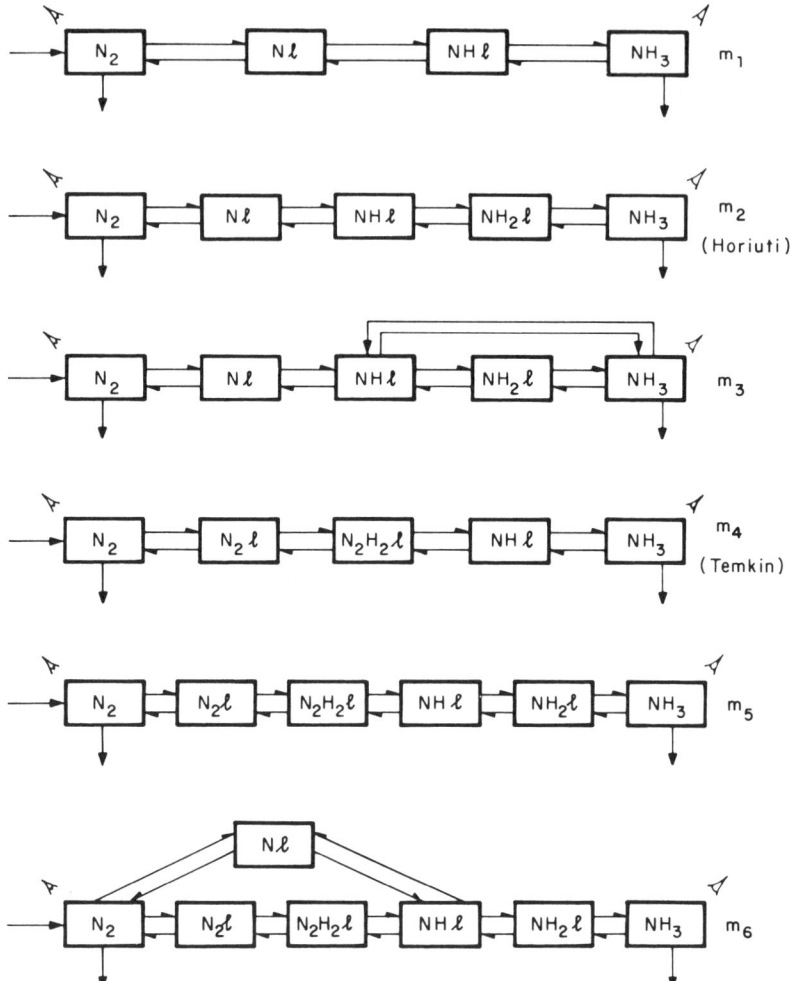

FIG. 6.3 Models for nitrogen tracing of ammonia synthesis, Example 6.4.1.

marking in nitrogen z_N is observed. Note that the models are all catenary or have modified catenary paths with bridging over some of the steps.

Table 6.2 summarizes the results for the six models given in Fig. 6.3 by following the transfer function method. Details of procedure are given in the supplementary material by Happel *et al.* (1986).

Conclusions regarding identifiability are given on the principal diagonal in Table 6.2. For the ith case, sgi indicates that the model is structurally globally identifiable. The designation "?" for model 6 with a bridged structure

6.4 General Procedure for Model Testing

TABLE 6.1 Mechanisms for Ammonia Synthesis

Mechanisms	$s_1 + s_2 + s_3$	$s_4 + 2s_5$	$s_6 + s_9$	s_7	s_8
m_1	0	1	0	2	1
m_2	0	1	2	0	3
m_3	0	1	-1	3	0
m_4	1	0	0	2	0
m_5	1	0	2	0	2
m_6	3	-2	2	0	0

TABLE 6.2 Identifiability and Distinguishability of Ammonia Synthesis Models

Models	$\mathcal{M}_1$	$\mathcal{M}_2$	$\mathcal{M}_3$	$\mathcal{M}_4$	$\mathcal{M}_5$	$\mathcal{M}_6$
$\hat{\mathcal{M}}_1$	sgi	sd	sd	sd	sd	sd
$\hat{\mathcal{M}}_2$	sd	sgi	sd	$\overline{sd}$	sd	sd
$\hat{\mathcal{M}}_3$	$\overline{sd}$	$\overline{sd}$	sgi	$\overline{sd}$	sd	sd
$\hat{\mathcal{M}}_4$	sd	$\overline{sd}$	sd	sgi	sd	sd
$\hat{\mathcal{M}}_5$	sd	sd	sd	sd	sgi	sd
$\hat{\mathcal{M}}_6$	sd	sd	sd	sd	$\overline{sd}$	?

indicates that this model requires further study to determine whether it is identifiable. For special values of their parameters, m_3 reduces to m_1 and m_5 to m_2. If concentrations of the surface intermediates could be resolved by appropriate spectroscopy, it might be possible to model these more complicated cases.

The remainder of the table gives the properties of distinguishability of the models, taken by pairs. Thus sd in the position i,j (ith row, jth column) indicates that the model $\hat{\mathcal{M}}_i$ is structurally distinguishable from a "process" $\mathcal{M}_j$. The opposite is indicated by $\overline{sd}$.

Thus, for example, the entry $\overline{sd}$ corresponding to $\hat{\mathcal{M}}_6, \mathcal{M}_1$, means that, if a set of parameters can be found for the $\mathcal{M}_1$, it will also be possible to find a set of parameters that would satisfy $\hat{\mathcal{M}}_6$. This case of indistinguishability in which the number of compartments in one model ($\mathcal{M}_6$) exceeds that in the other ($\mathcal{M}_1$) simply involves forcing parameters of the more general model to equal zero. From a practical viewpoint, if these parameters are numerically negligible it would be reasonable to accept the simpler model.

A different kind of indistinguishability is that in which the same number of compartments is involved such as distinguishing between Horiuti's ($\mathscr{M}_2$) and Temkin's ($\mathscr{M}_4$) models. If nitrogen tracing were employed, experiments in which nitrogen isotope scrambling was tested

$$^{15}N^{15}N + {}^{14}N^{14}N \rightleftharpoons 2\,^{15}N^{15}N$$

could be employed to distinguish between these mechanisms.

Finally, of course, deuterium tracing could serve to model steps involving hydrogen adsorption, which are not modeled by the transients of nitrogen transfer.

6.5 Discussion

In this chapter two different but complementary techniques have been advanced with the objective of obtaining more fundamental information on mechanistic chemical parameters than is available by the usual methods.

The structure of transient tracing models can be examined mathematically, independent of numerical data, to consider the problems of global identifiability and distinguishability. Methods have been presented for determining whether the parameters can be uniquely determined and whether it is possible to distinguish models from each other.

For cases in which models are globally unidentifiable or indistinguishable a further step in model building would be to generate the set of local solutions consistent with given data and an appropriate model. From such local solutions it is then possible to generate sets of local solutions for corresponding indistinguishable models. Physical limitations posed by other experimental data and theoretical considerations can then be employed to narrow the choice of acceptable parameters and models. In the next chapter the use of these concepts in connection with tracer data will be illustrated.

List of Symbols

A	Matrix associated with states
A	Reactant molecule
B	Matrix associated with tracer input
B	Product molecule
C	Observation matrix
C^i	Concentration of terminal species (i = A or B)
C^{Al}	Concentration of intermediate on catalyst surface
F_F^A	Rate of feed of A to the system
F^A	Rate of removal of A from the system

F^B	Rate of removal of B from the system
$H(s, \theta), \hat{H}(s, \hat{\theta})$	Transfer function
$\mathbf{I}$	Identity matrix
$\mathcal{M}$	Model
s	Laplace transform variable
t	Time
$\mathbf{u}$	Vector of tracer input
$v_{\pm i}$	Velocity of mechanistic step $i = 1, 2$ per unit mass
W	Weight of catalyst
$\mathbf{y}$	Vector of observed tracer output fraction
$z_F^A = u$	Fractional marking of feed
z^A, z^{Al}, z^B	Fractional marking of A, Al, and B
$\mathbf{z}$	State vector
$\dot{\mathbf{z}}$	dz/dt
β	Dead volume of the system
θ	Parameter vector

References

Beck, J. V., and Arnold, K. J. (1977). "Parameter Estimation in Engineering and Science." Wiley, New York.
Bellman, R., and Åström, K. J. (1970). *Math. Biosci.* **7**, 329.
Happel, J. (1974). *J. Res. Inst. Catal. Hokkaido Univ.* **22**, 206.
Happel, J., and Sellers, P. H. (1982). *Ind. Eng. Chem. Fundam.* **21**, 67.
Happel, J., and Sellers, P. H. (1983). *Adv. Catal.* **32**, 273.
Happel, J., Cheh, H. Y., Otarod, M., Ozawa, S., Severdia, A. V., Yoshida, T., and Fthenakis, V. (1982). *J. Catal.* **72**, 314.
Happel, J., Walter, E., and Lecourtier, Y. (1986). To be published in *I&EC Fundam.*.
Horiuti, J. (1973). *Ann. N. Y. Acad. Sci.* **213**, 5.
Lecourtier, Y., and Walter, E. (1981). *IEEE Trans. Autom. Control* **AC-26**, 800.
Park, S. W., and Himmelblau, D. M. (1982). *Chem. Eng. J.* **25**, 163.
Temkin, M. I. (1979). *Adv. Catal.* **28**, 173.
Walter, E. (1982). "Identifiability of State Space Models—Lecture Notes in Biomathematics," Vol. 46. Springer-Verlag, Berlin and New York.
Walter, E., and Lecourtier, Y. (1982). "Mathematics and Computers in Simulation." XXIV, p. 472. North-Holland Publ., Amsterdam.
Walter, E., Lecourtier, Y., and Happel, J. (1984a). *IEEE Trans. Autom. Control.* **AC-29**, 56.
Walter, E., Lecourtier, Y., Raksanyi, A., and Happel, J. (1984b). *IMACS Int. Symp. Biomed. Syst. Model.*, 2nd Bethesda, August 6-10.
Walter, E., Lecourtier, Y., and Happel, J. (1986). To be published in *AIChE J.*, June 1986.

Chapter 7

Transient Tracing

7.1 Introduction

As discussed in Chapter 4, a considerable simplification in mathematical treatment is possible, if a rapidly recirculating or stirred-tank type of reactor is used to study reaction systems by transient tracing because concentration changes in space, which are present in tubular reactors, are avoided. In the studies reported in this chapter such reactors are employed. In order for the overall reaction to be at steady state, feed must be continuously introduced and product must be continuously withdrawn. When steady state is reached, a known concentration of tagged species is substituted for untagged species in one of the feed streams at concentration z_F and maintained at that concentration, producing a step function in tracer input. Other methods of introducing tracer, such as a pulse or a continuing function of time, are possible.

The fractional marking z of components in the circulating system and the product will change with time and may be expressed as material balances, one for each species present in both the gas and chemisorbed phases. These material balances, as discussed previously, may be written as in Eq. (4.7):

$$d\mathbf{z}/dt = \mathbf{A}\mathbf{z} + \mathbf{b} \qquad (7.1)$$

The more general form used in systems analysis, Eq. (6.1) is not usually needed in tracer studies. The column vector **b** is determined from the feed input of tracer, which is constant. The **A** matrix consists of coefficients involving the parameters in the material balances, including step velocities and surface concentrations of adsorbed species. The solution of Eq. (7.1) may be expressed as (Amundson, 1966)

$$\mathbf{z} = \sum_j C_j k_j e^{\lambda_j t} - \mathbf{A}^{-1}\mathbf{b} \qquad (7.2)$$

where the k_j are eigenvectors, the λ_j are eigenvalues of A, and the constants C_j are determined from the initial conditions at $t = 0$. It is, of course, possible to determine **z** if the **A** matrix is specified, but our problem is the *inverse*, as discussed in Chapter 4, namely to determine **A** with incomplete information for **z** since only terminal species markings are generally observed.

The usual procedure for testing models is to propose some minimal structure, assign reasonable initial values for the parameters, and by iteration obtain the computer solution for the parameters that best fits the observed data. Freedom in choosing a model with a specified number of compartments (Berman, 1971) is expressed in terms of a minimum number of variables equal to the difference between the number of model parameters (surface concentrations and step velocities) and the number of invariants. For a model to be identifiable the difference must be zero. Thus, the procedure employed assumes that sufficient information is available so that a minimal model can be found. The set of all possible models may be obtained by linear transformation from the generating model (Walter *et al.*, 1980).

Three possible situations exist, corresponding to

(1) a unique solution, which implies global identifiability;

(2) a finite number of solutions, which implies a set of models having their structure defined by appropriate values of the parameters but with outputs indistinguishable from the generating model; this is described as local identifiability; and

(3) an infinite number of solutions; the structure is not identifiable and the model set obtained expresses the structural ambiguity of the data obtained.

In practice, the following problems may arise:

(1) the nature of the system studied is not sufficiently known;
(2) the sites available for measurement are too restricted;
(3) the times when measurements can be made are restricted;
(4) there are restrictions in concentrations, temperature, and pressure of operation;
(5) the data contain fluctuations due to instrumental noise;
(6) the system must be studied by various techniques that are related in a complicated way;

Our aim is to find the "decay constants" λ_j and the amplitudes A_j. An important problem in modeling centers around specifying the j compartments in Eq. 7.2. On the surface the problem seems simple, but the straightforward mathematical solution does not indicate the enormous practical difficulties that arise if we try to apply it to physical problems (Lanczos, 1961). Thus, if the separation of four or five exponentials is desired, an accuracy of six to eight significant figures is required in the observed data for successful solution. Even the separation of three exponentials corresponding to three compartments is difficult.

As soon as we proceed with the determination of parameters for more than one model, the additional problem of distinguishability arises. It is in the areas of identifiability and distinguishability that it is important to rely on physical data regarding the presence of surface intermediates.

All these factors serve to emphasize the fact that it is not desirable to employ more complicated models than those required to correlate the observed data. The matter of "goodness-of-fit" is most important in judging the results of curve fitting techniques discussed in the following section.

7.2 Generalized Data Fitting

A general procedure for obtaining parameters corresponding to phenomena described by sets of ordinary first-order differential equations has been developed by Spencer and forms the basis of the discussion in this section (Spencer et al., 1971; Happel et al., 1977). In systems of this type we define "state variables" as those variables that describe the state of the system at any time. The ordinary differential equations that describe the evolution of the state variables in time are called the "system equations." In our case, the state variables refer to tracer fractional concentrations z if Eq. (7.1) is to be solved. A more detailed discussion of the subject of curve fitting is given in the book by Bard (1974).

The problem for solution by computer can be formulated in terms of the following system equations

$$dy_i/dt = g_i(y_n, b_j)$$
$$y_i(0) = a_i$$
$$i, n = 1, 2, \ldots, NS$$
$$j = 1, 2, \ldots, NP$$

(7.3)

Here t is the independent variable, y_i is the ith state variable, b_j is the jth parameters, and a_i is the initial value of g_i. The functions g_i can be linear or nonlinear. In Eq. (7.3) NS represents the number of state variables and NP represents the number of parameters to be determined. We assume that the mth state variable y_m has been measured at times t_k; $k = 1, 2, \ldots, NX$ and

7.2 Generalized Data Fitting

denote

$$y_m(t_k) \triangleq y_k, \quad k = 1, 2, \ldots, NX \tag{7.4}$$

Here y_k is a function of the unknown parameters b_j

$$y_k = y_k(b_j) \tag{7.5}$$

and the measured value of y_m at time t_k is denoted by y_k^*. The problem is to find parameter values b_j, such that the NX equations,

$$y_k(b_j) - y_k^* = 0 \tag{7.6}$$

are satisfied. If $NX = NP$, we have NP equations in NP unknowns, which could be solved iteratively by the Newton–Raphson method. But generally there will be more experimental values than parameters, i.e., $NX > NP$. In this case we cannot demand an exact fit to the data, but must be satisfied with a "best" fit. In the present development the "best" fit is taken as a least-squares fit that corresponds to minimizing the objective function S,

$$S(b_j) \triangleq \sum_{k=1}^{NX} (y_k^* - y_k)^2 \tag{7.7}$$

The problem is solved iteratively. The computer program proceeds by guessing values for the parameters b_j, using parameter increments to calculate new and improved values, and finally at convergence arrives at a least-squares fit. If small enough parameter increments are chosen, the system can be linearized by a Taylor series expansion and if second- and higher-order terms are omitted, the following is obtained

$$y_k(b_j + \Delta b_j) \cong y_k(b_j) + \sum_{j=1}^{NP} \frac{dy_k}{db_j} \Delta b_j \tag{7.8}$$

$$k = 1, 2, \ldots, NX, \quad j = 1, 2, \ldots, NP$$

We would like to find parameter increments b_j such that

$$y_k(b_j + \Delta b_j) = y_k^* \tag{7.9}$$

This corresponds to the equation

$$\sum_{j=1}^{NP} \frac{\partial y_k}{\partial b_j} \Delta b_j = y_k^* - y_k \triangleq E_k \tag{7.10}$$

where E_k is the error criterion for judging when convergence is achieved. Often E_k is taken as 1% of y_k^*, which provides a satisfactory choice consistent with the noise level of the tracer data. Equation (7.10) may be written in matrix–vector notation as

$$\mathbf{B} \, \Delta \mathbf{b} = \mathbf{e} \tag{7.11}$$

where by definition

$$B_{kj} = \frac{\partial y_k}{\partial b_j}, \quad \mathbf{e} \triangleq \begin{bmatrix} y_1^* - y_1 \\ \vdots \\ y_m^* - y_m \end{bmatrix}, \quad \Delta \mathbf{b} \triangleq \begin{bmatrix} b_1 \\ \vdots \\ b_{NP} \end{bmatrix}$$

where B is an NX by NP matrix in which each column can be considered as an NX-dimensional vector, called a curve-increment vector, since it represents the direction (in NX-dimensional space) that a unit change in the corresponding parameter produces in the y_m curve. For example, when $NX = 3$ and $NP = 2$, $\mathbf{B}\,\Delta\mathbf{b} = \mathbf{e}$ has the form

$$\begin{bmatrix} \frac{\partial y_1}{\partial b_1} & \frac{\partial y_1}{\partial b_2} \\ \frac{\partial y_2}{\partial b_1} & \frac{\partial y_2}{\partial b_2} \\ \frac{\partial y_3}{\partial b_1} & \frac{\partial y_3}{\partial b_2} \end{bmatrix} \begin{bmatrix} \Delta b_1 \\ \Delta b_2 \end{bmatrix} = \begin{bmatrix} y_1^* - y_1 \\ y_2^* - y_2 \\ y_3^* - y_3 \end{bmatrix} \triangleq \begin{bmatrix} E_1 \\ E_2 \\ E_3 \end{bmatrix}$$

If we call the curve-increment vectors F_j

$$F_j = \begin{bmatrix} \frac{\partial y_1}{\partial b_j} \\ \vdots \\ \frac{\partial y_{NX}}{\partial b_j} \end{bmatrix} \tag{7.12}$$

then the problem is to express the error vector $\mathbf{e}$ as a linear combination of the F_j, i.e.,

$$\mathbf{e} = \Delta b_1 F_1 + \Delta b_2 F_2 + \cdots + \Delta b_{NP} F_{NP} \tag{7.13}$$

We would like to make the error vector equal to zero, but, in general, there will not be an exact solution to Eq. (7.11). We can only ask for a linear combination of F_j values that will be as close as possible to $\mathbf{e}$. Thus, we look for parameter increments that will minimize the function S, in which

$$S(\Delta b_j) = \sum_{k=1}^{NX} \left(\sum_{j=1}^{NP} \frac{\partial y_k}{\partial b_j} \Delta b_j - E_k \right)^2 \tag{7.14}$$

A necessary condition for S to be a minimum is that

$$\partial S / \partial b_l = 0$$

We designate this as $(SS)_{\min}$, and, in general, it will correspond to

$$\frac{\partial S}{\partial \Delta b_l} = 2 \sum_{k=1}^{NX} \left(\sum_{j=1}^{NP} \frac{\partial y_k}{\partial b_j} \Delta b_j - E_k \right) \frac{\partial y_k}{\partial b_l} = 0 \tag{7.15}$$

7.2 Generalized Data Fitting

By changing the order of summation, we have the normal equations

$$\sum_{j=1}^{NP} \left(\sum_{k=1}^{NX} \frac{\partial y_k}{\partial b_j} \frac{\partial y_k}{\partial b_l} \right) \Delta b_j = \sum_{k=1}^{NX} \frac{\partial y_k}{\partial b_l} E_k, \quad l = 1, 2, \ldots, NP. \quad (7.16)$$

These NP equations in NP unknowns, Δb_j, determine parameter increments that minimize S, provided that initial parameter guesses are in the vicinity of a minimum. If not, a different initial choice is necessary. In matrix notation

$$\mathbf{B}^T \mathbf{B} \, \Delta \mathbf{b} = \mathbf{B}^T \mathbf{e} \quad (7.17)$$

where $\mathbf{B}^T$ is the transpose of $\mathbf{B}$. The term $\mathbf{B}^T \mathbf{B}$ is symmetric, and the normal equations will always be compatible, although $\mathbf{B}^T \mathbf{B}$ can be singular. The increment $\Delta \mathbf{b}$ is found by multiplying both sides of Eq. (7.17) by the inverse of $\mathbf{B}^T \mathbf{B}$ to give

$$\Delta \mathbf{b} = (\mathbf{B}^T \mathbf{B})^{-1} \mathbf{B}^T (\mathbf{e}) \quad (7.18)$$

These "normal" equations refer to the Gauss–Newton approach, which proceeds by guessing values for the parameters b_j, calculating the matrix $\mathbf{B}$ and the vector $\mathbf{e} = y_k^* - y_k$, solving Eq. (7.18) and finally using the parameter increments Δb_j to calculate new and improved parameter values $b_j = b_j + \Delta b_j$. This process is repeated until $\mathbf{e}$ is reduced to its minimum length, which finally corresponds to a least-squares fit. At convergence, $\mathbf{B}^T \mathbf{e} = 0$, i.e., indicated values of $\Delta b_j = 0$.

A drawback to this straightforward approach is that for initial guesses of the parameters that are not close to the true value, the method will often diverge. If the Gauss–Newton initial guess is far from the true value, then rapid convergence results. Another popular approach is the steepest descent algorithm. It is analogous to the Newton–Raphson method, but extended to higher vector spaces. The steepest descent method takes small steps, and will always finally converge to the minimum, if it exists. A drawback is that, if one is far from the true parameter values in the initial guess, many iterations must be employed with resulting large computer time usage.

The Marquardt (1963) method takes advantage of both the preceding methods. The following is a synopsis of its derivation. Let S be a diagonal scaling matrix. Then we can multiply both sides of Eq. (7.17) as follows:

$$\mathbf{S}^T \mathbf{B}^T \mathbf{B} \mathbf{S} \mathbf{S}^{-1} \, \Delta \mathbf{b} = \mathbf{S}^T \mathbf{B}^T \mathbf{e} \quad (7.19)$$

where

$$S_{jj} = 1/[(\mathbf{B}^T \mathbf{B})_{jj}]^{1/2}, \quad S_{jj}^{-1} = [(\mathbf{B}^T \mathbf{B})_{jj}]^{1/2}$$

Thus, the diagonal elements are equal to 1. Now the scaled normal equations are modified by adding the Lagrange multiplier λ as shown:

$$(\mathbf{S}^T \mathbf{B}^T \mathbf{B} \mathbf{S} + \lambda \mathbf{I}) \mathbf{S}^{-1} \, \Delta \mathbf{b} = \mathbf{S}^T \mathbf{B} \mathbf{e} \quad (7.20)$$

where $\mathbf{I}$ is the identity matrix.

The Lagrange multiplier will act as a parameter to promote convergence. After multiplying by the inverse of the scaling matrix and rearranging, the following form is obtained:

$$[\mathbf{B}^T\mathbf{B} + \lambda \mathbf{S}^{-1}\mathbf{S}^{-1}]\Delta b = \mathbf{B}^T\mathbf{e} \quad (7.21)$$

and

$$\Delta b = [\mathbf{B}^T\mathbf{B} + \lambda \mathbf{S}^{-1}\mathbf{S}^{-1}]^{-1}\mathbf{B}^T\mathbf{e} \quad (7.22)$$

The effect of this modification is that the diagonal elements of the $\mathbf{B}^T\mathbf{B}$ matrix are increased by a factor of $(1 + \lambda)$. As λ is increased, the parameter increment is decreased, and the method becomes in the limit that of steepest descent

$$\Delta \mathbf{b} = (1/\lambda)\mathbf{B}^T\mathbf{e} \quad (7.23)$$

If λ is decreased, the method becomes, in the limit, that of the previously discussed Gauss–Newton procedure. The Marquardt algorithm essentially consists of a choice of values of λ for succeeding iterations. If the objective function shows an improvement over the previous iteration, then λ is halved. If, however, an improvement is not observed, λ is quadrupled. The former action results in taking larger steps, cutting down the number of iterations near the minimum. The latter will approach the steepest descent method in an attempt to move in the direction of the minimum. This procedure will finally result as before in convergence with $\mathbf{B}^T\mathbf{e} = 0$.

Although the determination of the best values of the parameters is of some value, it is desirable to have some information as to how reliable they are in order to compare results from one model with those from another. In order to accomplish this the error vector $\mathbf{e}$ is found by integrating, usually numerically by a Runge–Kutta method, the system equations to determine $y_k = y_m(t_k)$. The elements of the matrix $\mathbf{B}$ can be found by integrating a set of differential equations, called the variational equations, which are derived by partially differentiating each system equation with respect to each parameter. The variational equations have the form

$$\frac{d}{dt}\frac{\partial y_m}{\partial b_j} = \frac{\partial g_m}{\partial b_j}\bigg|_{y=\text{const}} + \sum_{n=1}^{NS} \frac{\partial g_m}{\partial y_n}\frac{\partial y_n}{\partial b_j} \quad (7.24)$$

where $j = 1, 2, \ldots, NP$ and $m, n = 1, 2, \ldots, NS$. This is a set of $NS \times NP$ simultaneous ordinary differential equations, which can be integrated along with the system equations to determine $(\partial y_m / \partial b_j)$ as functions of time. Then $\mathbf{B}$ is determined by

$$B_{kj} = \frac{\partial y_k}{\partial b_j} = \frac{\partial y_m}{\partial b_j}\bigg|_{t=t_k}$$

7.2 Generalized Data Fitting

The initial conditions for the variational equations are as follows

(1) if the jth parameter is *not* related to the initial value of y_m, then

$$\left.\frac{\partial g_m}{\partial b_j}\right|_{t=0} = 0$$

since the values of y_m at time zero depend only on the initial condition.

(2) if b_j is the initial value of g_m, i.e., $y_m(0) = b_l$, then

$$\left.\frac{\partial y_m}{\partial b_j}\right|_{t=0} = 1$$

In general,

$$\left.\frac{\partial y_m}{\partial b_j}\right|_{t=0} = \frac{\partial a_m}{\partial b_j}$$

Then we have all the initial conditions needed to simultaneously integrate the system equations and the variational equations. Figure 7.1 gives a schematic diagram for how this is accomplished in a computer program.

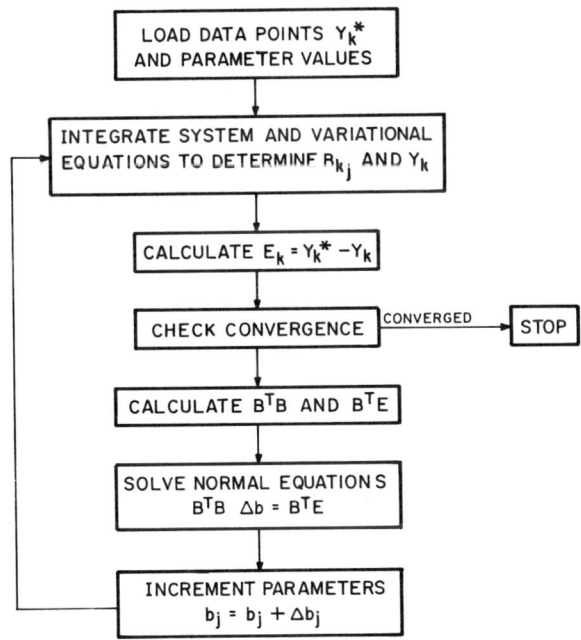

FIG. 7.1   Iterative procedure.

When the best parameter values have been obtained we have the condition

$$\sum_{k=1}^{NX} \frac{\partial y_k}{\partial b_j}(y_k^* - y_k) = 0 \qquad (7.25)$$

corresponding to $\mathbf{B}^T\mathbf{e} = 0$ or $F^j \cdot \mathbf{e} = 0$. This satisfies the geometrical condition that the residual vector $\mathbf{e} = y_k^* - y_k$ is orthogonal to each sensitivity vector F^j, where $F^j = \partial y/\partial b_j$.

Thus a small change in the jth parameter will not produce motion in the direction of reducing E_j, and the length of $\mathbf{e}$ is stationary with respect to each parameter. This corresponds to a minimum in the sum of the squares of residuals, since the sum of squares is proportional to $|e|^2$.

In modeling a set of proposed mechanisms, the initial choice is, of course, directed to the model corresponding to the smallest standard deviation s, which is designated as

$$\text{STDVN} = s = [SS_{\min}/(NX - NP)]^{1/2} \qquad (7.26)$$

The standard deviation obtained for the values predicted from tracer output should approach that corresponding to the accuracy of the determinations themselves, if the best model available is to be considered acceptable.

Models should be further examined by nonparametric tests. Lack of fit due to an incorrect or oversimplified model may be revealed by a nonrandom distribution of residuals (Siegel, 1956) using a runs test. A run is defined as a group of one or more consecutive residuals of the same sign. Usually the number of runs will lie between 3 and 20 for modeling such as we have conducted. A very large number of runs may indicate some type of cyclic disturbance. On the other hand, with less than three runs it is likely that too simple a correlation is being employed.

It is next desirable to consider the reliability of the parameter values determined by this procedure. To answer this question a statistical procedure is employed (Draper and Smith, 1968). If the model chosen is too complex, there may be many widely different sets of parameter values that will give equally good fits to the data. This can occur when the members of a group of parameters change simultaneously in such a way that the measured tracer values remain essentially constant. The parameters in the group are then said to be correlated, and the result is that only certain parameters can be determined accurately, while other parameters will be known only approximately or not at all.

We wish to trace the effect of small changes in the data $\Delta \mathbf{y}^*$, on the parameters $\hat{\mathbf{b}}$. If $\mathbf{y}^*$ is replaced by $\mathbf{y}^* + \Delta y$, b will be replaced by $\hat{\mathbf{b}} + \Delta \mathbf{b}$. Thus, we will have

$$\mathbf{B}^T(\mathbf{y}^* + \Delta \mathbf{y}^*) - (\mathbf{y} + \Delta \mathbf{y}) = 0 \qquad (7.27)$$

7.2 Generalized Data Fitting

At convergence, where $\Delta \mathbf{y}$ is the change in $\mathbf{y}$ due to the change in $\mathbf{b}$, we have

$$\Delta y_k = \sum_{j=1}^{NP} \frac{\partial y_k}{\partial b_j} \Delta b_j \quad \text{or} \quad \Delta \mathbf{y} = \mathbf{B}\, \Delta \mathbf{b}$$

Since $\mathbf{B}^T(\mathbf{y}^* - \mathbf{y}) = 0$, we have

$$\mathbf{B}^T(\mathbf{y}^* - \mathbf{y}) = \mathbf{B}^T \Delta \mathbf{y}^* - \mathbf{B}^T \mathbf{B}\, \Delta \mathbf{b} = 0$$

or

$$\mathbf{B}^T \mathbf{B}\, \Delta \mathbf{b} = \mathbf{B}^T \Delta \mathbf{y}^* \quad (7.28)$$

or

$$\Delta \mathbf{b} = (\mathbf{B}^T \mathbf{B})^{-1} \mathbf{B}^T \Delta \mathbf{y}^*$$

Equation (7.28) can be used to calculate $\Delta \mathbf{b}$ for $\Delta \mathbf{y}^*$, assuming that $\mathbf{B}$ does not change with small changes in $\mathbf{b}$. (Note the similarity to the normal equations $\mathbf{B}^T \mathbf{B}\, \Delta \mathbf{b} = \mathbf{B}^T \mathbf{e}$. As a result the process of determining $\hat{\mathbf{b}}$ gives exactly the same matrices, $\mathbf{B}^T \mathbf{B}$ and $\mathbf{B}$, that are needed to evaluate $\Delta \mathbf{b}$ from $\Delta \mathbf{y}^*$.) Consider next the expected value of $\Delta \mathbf{b}\, \Delta \mathbf{b}^T$. The variance–covariance matrix for parameter estimates is given by

$$\mathbf{V} \triangleq \mathbf{e}(\Delta \mathbf{b}\, \Delta \mathbf{b}^T) \quad (7.29)$$

For example in the case of $NP = 3$ this is:

$$\mathbf{V} \triangleq \mathbf{e}\{\Delta \mathbf{b}\, \Delta \mathbf{b}^T\} = \mathbf{e} \begin{bmatrix} \Delta b_1^2 & \Delta b_1\, \Delta b_2 & \Delta b_1\, \Delta b_3 \\ \Delta b_2\, \Delta b_1 & \Delta b_2^2 & \Delta b_2\, \Delta b_3 \\ \Delta b_3\, \Delta b_1 & \Delta b_3\, \Delta b_2 & \Delta b_3^2 \end{bmatrix}$$

The diagonal elements of $\mathbf{V}$ are the variances of the parameters and serve as a measure of the uncertainty in the parameter values. Using Eq. (7.28) in Eq. (7.29), we have

$$\mathbf{V} = \mathbf{e}((\mathbf{B}^T\mathbf{B})^{-1}\mathbf{B}^T(\Delta \mathbf{y}^*\, \Delta \mathbf{y}^{*T})\mathbf{B}(\mathbf{B}^T\mathbf{B})^{-1}) \quad (7.30)$$

Since $(\mathbf{B}^T\mathbf{B})^T = \mathbf{B}^T\mathbf{B}$, the matrix $\mathbf{e}\{\Delta \mathbf{y}^*\, \Delta \mathbf{y}^{*T}\}$ is the variance–covariance matrix of the measurements and is written as

$$\mathbf{U} = \mathbf{e}\{\Delta \mathbf{y}^*\, \Delta \mathbf{y}^{*T}\} = \begin{bmatrix} \mathbf{e}\{\Delta y_1^2\} & \mathbf{e}\{\Delta y_1\, \Delta y_2\} & \mathbf{e}\{\Delta y_1\, \Delta y_3\} \\ \mathbf{e}\{\Delta y_2\, \Delta y_1\} & \mathbf{e}\{\Delta y_2^2\} & \mathbf{e}\{\Delta y_2\, \Delta y_3\} \\ \mathbf{e}\{\Delta y_3\, \Delta y_1\} & \mathbf{e}\{\Delta y_3\, \Delta y_2\} & \mathbf{e}\{\Delta y_3^2\} \end{bmatrix}$$

The diagonal elements of $\mathbf{U}$ are the variances of the measured data, and the off-diagonal elements are the covariances. To interpret the data the simplest initial assumption is that

$$\mathbf{e}\{\Delta y_j\, \Delta y_k\} = \begin{cases} \sigma^2, & j = k \\ 0, & j = k \end{cases} \quad (7.31)$$

That is, we assume that each measurement has the same variance σ^2 and that the measurements are (reflecting only measurement errors and not lack of fit) not correlated. With this assumption, we have

$$\mathbf{U} = \sigma^2 \mathbf{I} \tag{7.32}$$

and

$$\mathbf{V} = \sigma^2 (\mathbf{B}^T \mathbf{B}^{-1})(\mathbf{B}^T \mathbf{B})(\mathbf{B}^T \mathbf{B}) \tag{7.33}$$

or

$$\mathbf{V} = \sigma^2 (\mathbf{B}^T \mathbf{B}^{-1}) \tag{7.34}$$

where $\mathbf{V}$ is the variance–covariance matrix. Here $\{(\text{STDVN})^2 \times C_{kk}\}$ is the kkth element of the matrix. In the computer program the SIGMA value σ_k associated with the estimation of the kth parameter is

$$\{(\text{STDVN})^2 \times C_{kk}\}^{1/2}$$

which may be designated as the computed estimate of standard error.

As discussed by Spencer et al. (1971) and Box (1960), it is possible to use these standard errors to calculate confidence limits for the values of the vector of b_k parameters. The idea behind confidence limit calculations is that there is a region in parameter space around the best parameter estimates b_k that contains, with a certain probability (say 95%), the true parameter values. For our purposes the confidence limit (CL) will be given by

$$\text{CL} = \{NP \times F_\alpha(NP, NX - NP)\}^{1/2} \{(\text{STDVN})^2 \times C_{kk}\}^{1/2} \tag{7.35}$$

Note that STDVN is given by Eq. (7.26). The F_α is the F-distribution statistic termed the variance ratio. Thus for a 95% confidence limit, we employ $F_{0.95}(NP, NX - NP)$.

At times it may happen that the standard deviation STDVN is too small to produce suitable values in the σ_k. In that case it is convenient to multiply C_{kk} by unity instead of $(\text{STDVN})^2$.

A convenient tool to employ in looking at the parameters simultaneously to determine whether they are independent is the correlation matrix. In this case the original $\mathbf{B}^T \mathbf{B}$ matrix is transformed to a scaled form in which all the diagonal elements are unity. The off-diagonal elements give a correlation coefficient between parameters. Theoretically they will be zero, if no correlation exists between coefficients. Thus, for example, if we are dealing with three parameters b_1, b_2, and b_3, the correlation matrix will look like

$$\begin{bmatrix} 1 & b_{12} & b_{13} \\ b_{21} & 1 & b_{23} \\ b_{31} & b_{32} & 1 \end{bmatrix}$$

The coefficients $b_{ij} = b_{ji}$ indicate the degree of correlation between parameters. Thus, if $b_{12} = b_{21}$ were less than unity, it would indicate some independence of parameters b_1 and b_2.

The application of these concepts will be illustrated by numerical examples in the following section as applied to models in which linear differential equations with constant coefficients are involved. In a final section special techniques for more complicated cases are considered.

A problem not considered in this book is that of weighting certain data, depending on the accuracy with which they are determined. Generally we have assumed that all tracer concentrations are likely to be measured with the same accuracy.

Another subject not treated here is that of experimental designs to obtain data more effectively. In our studies step functions of input have been employed for the most part, but it is possible that other forms of input would be attractive. Spencer *et al.* (1981) have also suggested a method for determining the effect of adding additional data points at certain stages during transient development of tracer concentration. The availability of an additional data point will cause the standard error of the *m*th parameter to be reduced, and the amount of this reduction is a measure of the "value" of the additional determination. In the absence of parameter correlation, this method shows the advantage of taking additional data points at times where the profile is most sensitive to the parameter in question.

To deal with complex systems it is essential to use a modern digital computer. A number of nonlinear regression computer programs are available from various sources. Perhaps the most elaborate is the SAAM (simulation analysis and modeling) program, which is available in codes for several larger computers (Berman and Weiss, 1978).

7.3 Typical Tracer Studies

In what follows we will illustrate by several examples of increasing complexity how the methods are applied in practice to modeling of heterogeneous catalytic reactions. These illustrations also serve as basis for organizing the content of pertinent studies that have been conducted in this field.

Example 7.3.1 Oxidation of Carbon Monoxide

The oxidation of carbon monoxide was studied in a gradientless recirculating reactor system at room temperature and atmospheric pressure over a commercial Hopcalite catalyst (Mine Safety Appliances Company, containing approximately 78.3 wt.% manganese dioxide and 13.1 wt.% cupric oxide)

(Happel et al., 1977; Kiang, 1978). It was found that the reactions could be modeled by the following three-step mechanism:

$$2CO + O_2 = 2CO_2$$

(i) $\quad O_2(g) + 2l \xrightarrow{v_{+1}} 2Ol$ (7.36)

(ii) $\quad CO(g) + Ol \xrightarrow{v_{+2}} CO_2 l$

(iii) $\quad CO_2 l \underset{v_{-3}}{\overset{v_{+3}}{\rightleftharpoons}} CO_2(g) + l$

where l indicates adsorption sites. It was determined, by marking of the product CO_2 with $^{13}CO_2$ and observing that no transfer of tracer to ^{13}CO occurred, that the first two steps are unidirectional. Further transient tracing experiments were conducted using gases containing ^{13}C and ^{18}O stable isotopes.

Details are given for a typical experiment in which ^{13}CO is marked to generate a step function. The compartmental model for the system consists of a three-compartment sequence. As shown in the previous chapter, this three-compartment model is structurally globally identifiable (sgi).

The equations corresponding to carbon atom balances for each of the three compartments—gaseous CO, adsorbed CO_2, and gaseous CO_2—involved in the carbon tracing are

$$\frac{F_F^{CO}}{W} z_F^{CO} - \frac{F^{CO} z^{CO}}{W} = \frac{\beta}{W} C^{CO} \frac{dz^{CO}}{dt} + v_{+2} z^{CO}$$

$$C^{CO_2 l} \frac{dz^{CO_2 l}}{dt} = v_{+2} z^{CO} + v_{-3} z^{CO_2} - v_{+3} z^{CO_2 l} \quad (7.37)$$

$$\frac{F_F^{CO_2}}{W} z_F^{CO_2} \frac{F^{CO_2} z^{CO_2}}{W} = \frac{\beta}{W} C^{CO_2} \frac{dz^{CO_2}}{dt} - v_{+3} z^{CO_2 l} + v_{-3} z^{CO_2}$$

where C^i is the concentration of species i ($i = CO, CO_2$) in the gas phase of the reaction system (g-moles/ml) (STP); $C^{CO_2 l}$, the concentration of species $CO_2 l$ adsorbed on solid catalyst (g moles/g); t, the time (min); F^i, the inlet or outlet flow rate of species i ($i = CO, CO_2$) (g-moles/ml); v_{+i}, the unidirectional step velocities ($i = 1, 2, 3$) (g-moles/min-g_{cat}); V, the overall reaction rate (referred to reactions as written, equivalent to rate of oxygen converted) (g-moles/min-g_{cat}); W, the weight of catalyst in the system (g); and z^i, the fraction of component i ($i = CO, CO_2, CO_2$) that contains tracer, in this case ^{13}C. Greek letter β is the volume of dead space, including that in the catalyst pores and apparatus (ml), and subscript F is feed stream.

For experiments in which there is a step-function input of ^{13}CO introduced after steady state with the same mixture containing unmarked species

7.3 Typical Tracer Studies

we have

State variables: $z^{CO}, z^{CO_2}, z^{CO_2}$
Measured variables: z^{CO}, z^{CO_2}
Parameters: C^{CO_2}, v_{+3}

Note that v_{-3} can be obtained by the equation $v_{-3} = v_{+3} - 2V$.

For convenience in computer programming the variables and constants are coded as follows:

State variables: $y_1 = z^{CO}$ Parameters: $b_1 = C^{CO_2}, C^{CO_2 l}$

$y_2 = z^{CO_2}$ $b_2 = v_{+3}$

$y_3 = z^{CO_2}$

Experimental constants:

$C_1 = \beta/W$, $C_2 = F^{CO}/W$, $C_3 = F^{CO_2}/W$, $C_4 = C^{CO}$

$C_5 = C^{CO_2}$, $C_6 = V$, $C_7 = (C_2 + 2C_6)/C_1 C_4$

$C_8 = 1/C_1 C_5$, $C_9 = C_3 - 2C_6$, $C_{10} = F_{in}^{CO_2}/W$,

$C_{11} = F_{in}^{CO}/W$, $C_{12} = z_{in}^{CO}$, $C_{13} = z_{in}^{CO_2}$

$C_{14} = C_{10} C_{12}/C_1 C_4$, $C_{15} = C_{11} C_{13}/C_1 C_5$

Systems equations:

$F(1) = dy_1/dt = C_{14} - C_7 y_1$

$F(2) = dy_2/dt = [2C_6 y_1 + (b_2 - 2C_6) y_3 - b_2 y_2]/b_1$

$F(3) = dy_3/dt = C_{15} + C_8 b_2 y_2 - C_8 (b_2 + C_9) y_3$

Variational equations:

$F(4) = d(dy_1/dt)db_1 = -C_7 y_4$

$F(5) = d(dy_1/dt)db_2 = -C_7 y_5$

$F(6) = d(dy_2/dt)db_1 = -F(2)/b_1 + [2C_6 y_4 - b_2 y_6 - (2C_6 - b_{2y}) y_8]/b_1$

$F(7) = d(dy_2/dt)db_2 = [2C_6 y_5 - y_2 - b_2 y_7 + y_3 - (2C_6 - b_2) y_9]/b_1$

$F(8) = d(dy_3/dt)db_1 = C_8 b_{2yy6} - C_8 (C_9 + b_2) y_8$

$F(9) = C_8 y_2 + C_8 b_2 y_7 - C_8 y_3 - C_8 (C_9 + b_2) y_9$

where,

$y_4 = dy_1/db_1$, $y_5 = dy_1/db_2$, $y_6 = dy_2/db_1$

$y_7 = dy_2/db_2$, $y_8 = dy_3/db_1$, $y_9 = dy_3/db_2$

As indicated in the previous section, the computer program integrates the system and variational equations denoted by $F(1)$–$F(9)$ to determine B_{kj} and y_k. It then repeats the procedure by performing iterations to obtain the converged values of these coefficients.

In the present case, we see that since v_{-1} and v_{-2} have previously been determined to be zero, the first equation $F(1)$ does not contribute to parameter determinations. Its solution only serves to check the random error in marking that appears in the z^{CO} tracer effluent. Thus, only $F(2)$ and $F(3)$ contribute to parameter determination.

Details of a sample run number 091175, are given to show the type of output obtained. In this run, conducted at atmospheric pressure and 24°C, an oxidized catalyst was employed using a feed mixture containing a ratio of $CO/O_2/CO_2 = 6/3/1$. Flow rate of the feed gas was 2.00 ml/min (STP) and that of helium carrier gas was 100 ml/min. Catalyst weight employed was 0.25 g. Oxygen consumption was 0.18 ± 0.03 ml/min/g, corresponding to 7.5% combustion of the CO and O_2 product in the feed gas. The dead volume was measured independently at 90 ml.

These measurements correspond to the following values for the coded constants:

$C_1 = 360$, $\qquad C_5 = 0.00284$, $\qquad C_{12} = 0.1502$

$C_2 = 4.44$, $\qquad C_6 = 0.1788$, $\qquad C_{13} = 0.013$

$C_3 = 1.16$ $\qquad C_{10} = 4.8$,

$C_4 = 0.01089$, $\qquad C_{11} = 0.8$,

The computer output using an error criterion $E_k = 0.0001$ was as follows for z^{CO_2}

Time (min)	Experimental	Calculated	Residual
0	0.013		
3	0.02916	0.03104	−0.00187
5	0.04481	0.04177	0.00304
7	0.04826	0.04788	0.00038
8	0.04997	0.04982	0.00015
10	0.05027	0.05232	−0.00205
15	0.05493	0.05469	0.00024
20	0.05344	0.05522	−0.00178
25	0.05621	0.05534	0.00097
30	0.05414	0.05537	−0.00122
35	0.05333	0.05537	−0.00204

In this run the points were taken more frequently at initial times in order to space them roughly equally over the curve of z^{CO} versus time. The initial marking observed is due to the natural abundance of ^{13}C in carbon monoxide.

The marking level of the ^{13}CO introduced as a step function was $z_F^{CO} = 0.15 \pm 0.008$, which would correspond to a steady-state concentration of $z^{CO_2} = 0.055$ after tracer admission. This value was essentially attained after 20 min.

Examination of these residuals shows seven runs, so the curve seems to represent the data adequately. The parameters with corresponding computed estimates of standard error were

$$C^{CO_2}(\text{ml/g}) = 2.36 \pm 0.25$$

$$v_{+3}(\text{ml CO}_2/\text{min/g}) = 2.39 \pm 0.10$$

The parameter correlation matrix is

$$\begin{bmatrix} 1 & 0.944 \\ 0.944 & 1 \end{bmatrix}$$

indicating some correlation between the exchange velocity and surface CO_2 concentration. The standard error in C^{CO_2} is also somewhat larger than the accuracy of reaction rate determination. However, the lack of trend in the output marking indicates that a more accurate determination of z^{CO_2} would be desirable. Experiments at a higher marking level could accomplish this and perhaps support a more sophisticated model. However, the evidence is clear that a substantial concentration of chemisorbed or combined CO_2 is present on the catalyst.

In the original publications cited in this example, a more complicated model in which CO_2 can be formed on two different types of sites was also considered, but the computer result indicated that such a mechanism was unlikely. Another earlier study conducted by us was also cited in which $^{13}CO_2$ marking was used, and the method of moments was used to interpret the data. The method of moments enables an explicit relationship to be obtained for v_{-3}. However, when applied to more complicated cases it does not have the versatility of the direct computer modeling of the differential equations involved. Transient tracing of the CO oxidation reaction using ^{18}O marked O_2 and CO_2 was also attempted but interpretation of the results is more difficult due to exchange of marked oxygen with lattice oxygen in the catalyst.

Another series of tests was conducted on the methanation of carbon monoxide by hydrogenation using a commercial nickel catalyst (Harshaw–104T, 60 wt. % nickel on silica) in a gradientless recirculating reactor with ^{13}CO transient tracing Fthenakis, 1978; Happel et al., 1980, 1981; Otarod et al., 1983). The following example illustrates the procedure used, in this case without knowledge of whether the system was structurally globally identifiable.

Example 7.3.2 Methanation Using ^{13}CO

Several different mechanisms were considered for modeling the transient tracing of methanation. Depending on the choice of elementary steps, as discussed in Chapter 3, a number of mechanisms can be considered possible. To narrow down the possibilities, it was necessary to consider results from previous investigations, and it was decided to eliminate oxygen-containing intermediates in modeling. Of special importance were separate studies using deuterium tracing that enabled the relative proportions of carbonaceous species to be assessed (Happel et al., 1982). Deuterium tracing had shown that concentrations of CH_2l and CH_3l were much smaller than that of CHl so it was decided to lump these species as CH_xl since they could not be uniquely identified. The lumped parameter CH_xl may be slightly larger than the sum of CHl, CH_2l, and CH_3l concentrations since a sequence of compartments will contribute greater delay in appearance of marking in CH_4 than a single compartment with the same total concentration. Deuterium experiments also showed that the active "carbidic" carbon C^al was much less than that of inactive "graphitic" carbon C^*l and hydrocarbon CH_xl species. Computer modeling was conducted over a spectrum of initial assumed values of the parameters within the limitations posed by these previous studies and resulted in a solution. From the standpoint of identifiability, restrictions based on independent information are necessary, as discussed in the next example that considers the structures of two possible models.

The model for carbon tracing used in this example is designated as model 1 and can be written as

$$CO + l \underset{}{\overset{v_1}{\rightleftharpoons}} COl$$

$$COl + l \overset{v_2}{\longrightarrow} C^a + O$$

$$C^al \underset{}{\overset{v_3}{\rightleftharpoons}} C^*l$$

$$C^*l + Hl \overset{v_4}{\longrightarrow} CHl + l \qquad (7.38)$$

$$CHl + Hl \overset{v_5}{\longrightarrow} CH_2l + l$$

$$CH_2l + Hl \overset{v_6}{\longrightarrow} CH_3l + l$$

$$CH_3l + Hl \overset{V}{\longrightarrow} CH_4 + 2l$$

The first step is taken to be at equilibrium and the step velocities are rapid so they cannot be determined. The second step is unidirectional. The third step, exchange between active and inactive carbon, is at equilibrium but need not be rapid since this exchange v_3 is not related to the methane formation rate. The remaining steps are taken as undirectional and via atomic hydrogen. The model, as shown, would require determination of the following seven parameters: $C^{COl}, C^{C^al}, C^{C^*l}, C^{CHl}, C^{CH_2l}, C^{CH_3l}$, and the velocity v_3. The three C^{CHl}, C^{CH_2l}, and C^{CH_3l} species are combined, as explained previously.

Material balances for the mechanisms for ^{13}C tracing, leaving out balances

7.3 Typical Tracer Studies

for C^{CO_l}, which is determined separately, are

$$dz^{C_{al}}/dt = (V/C^{C_{*l}})(z^{CO_l} - z^{C_{al}}) - (v_3/C^{C_{al}})(z^{C_{al}} - z^{C_{*l}})$$
$$dz^{C_{*l}}/dt = (v_3/C^{C_{al}})(z^{C_{al}} - z^{C_{*l}})$$
$$dz^{CH_xl}/dt = (V/C^{CH_xl})(z^{C_{al}} - z^{CH_xl}) \qquad (7.39)$$
$$dz^{CH_4}/dt = (VW/\beta C^{CH_xl})(z^{CH_xl} - z^{CH_4})$$

where C^il is the concentration of chemisorbed species, $i = C^a, C*, C^{CH_x}$ (ml/g) (STP); C^{CH_4}, the concentration of gas-phase CH_4, volume fraction; t, the time following initial change to ^{13}CO (min; V, the rate of production of methane (ml/min/g) (STP); $v_e = v_3$, the exchange velocity between active and inactive carbon, W, the weight of catalyst (g); β, the volume of dead space (ml); and z^{CO} and z^{CH_4} are observed as functions of time. Figure 7.2 shows the compartmental model schematically.

There are four parameters to be determined from the data $C^{C_{al}}, C^{C_{*l}}, C^{CH_xl}$, and v_e. A typical set of these parameters is listed as follows, using data reported by Otarod et al. (1983). The data correspond to run number 011680, but the parameters are not exactly the same as in the original paper because a slightly different method of computation was employed. The ^{13}CO data were used directly instead of using a correlation as had previously been done.

^{13}CO Step-up – 211°C	
Inlet flow rates (ml/min) NTP	
H_2	4.47
$^{13}CO + CO$	1.19
He	99.23
Inlet fractional marking	
z	19.27
Outlet flow rates (ml/min) NTP	
H_2	2.35
$^{13}CO + CO$	0.31
$^{13}CH_4 + CH_4$	0.80
$^{13}CO_2 + CO_2$	0.006
H_2O	0.88
Calculated parameters	
$C^{C_{al}}$ (ml/g/NTP)	1.630
$C^{C_{*l}}$ (ml/g/NTP)	5.338
C^{CH_xl} (ml/g/NTP)	0.164
v_e (ml/min/g, NTP)	0.177

Note: Catalyst wt. = 2.43 g; pressure = 1 atm; and dead space = 118 ml.

The solution obtained represents the data satisfactorially but the very low value for CH_xl is in disagreement with the deuterium experiments. However, this is only one of possible local solutions, as shown in the following example.

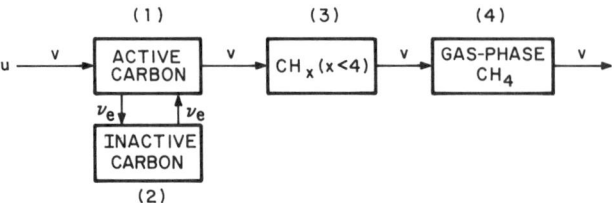

FIG. 7.2 First compartmental model for ^{13}C tracing of methanation.

Example 7.3.3 Identifiability in ^{13}CO Tracing of Methanation

Referring to Fig. 7.2 again, the compartments have the same designation as in the previous example. In order to simplify the appearance of the following derivations we rewrite the differential equations of Example 7.3.2 as

$$\begin{aligned}
C_1\dot{x}_1 &= -(V + v_e)x_1 + v_e x_2 + vu \\
C_2\dot{x}_2 &= -v_e x_2 + v_e x_1 \\
C_3\dot{x}_3 &= -V x_3 + V x_1 \\
C_4\dot{x}_4 &= -V x_4 + V x_3
\end{aligned} \quad (7.40)$$

The reactor being at steady state, C_i, V, and v_e are constants. In response to the input u, the state variables x_i vary according to a linear kinetics, even if the chemical reactions themselves are nonlinear.

The only quantities accessible for measurement in this experiment are the input $u(t)$ and the exit $y(t) = x_4(t)$, both measured for $t \in (t_1, t_2)$. The values of the constants C_4 and V are also known. The parameters that we seek to estimate by the measurements are C_1, C_2, C_3, and v_e.

The method employed for testing structural properties which is convenient for these systems is based on first obtaining the transfer function as discussed in Chapter 6 (see Walter, 1982). The transfer function $\mathbf{H}(s) = \mathbf{C}(s\mathbf{I} - \mathbf{A})^{-1}\mathbf{B}$ contains all the information regarding input–output behavior. In the case of a model structure $(A(\theta), B(\theta), C(\theta))$ coefficients of the transfer function depend on the parameters θ. We will call $\boldsymbol{\psi}(\theta)$ the vector formed from all of these coefficients in a fixed order.

For the model in Fig. 7.2 the transfer function can be obtained by hand or more readily by a program written in REDUCE (Hearn, 1983):

$$\mathbf{H}_1(s, \theta) = \frac{1}{1 + (C_4/V)s} \quad (7.41)$$

$$\times \frac{1 + (C_2/v_e)s}{\begin{aligned}1 &+ s[(C_2/v_e) + (C_1/V) + (C_2/V) + (C_3/V)] \\ &+ s^2[(C_1 C_2/V v_e) + (C_2 C_3/v_e V) + (C_1 C_3/V^2) + (C_2 C_3/V^2)] \\ &+ s^3[(C_1 C_2 C_3/V v_e V)]\end{aligned}}$$

7.3 Typical Tracer Studies

The first factor in this equation contains no parameters, so consideration can be limited to the second factor, which in order to simplify the notation, can be written with θ defined as

$$\theta_1 = C_1/V, \quad \theta_2 = C_2/V, \quad \theta_3 = C_3/V, \quad \theta_4 = C_2/v_e \quad (7.42)$$

Since V is known, there exists a one-to-one correspondence between the newly defined θ and the original parameters C_1, C_2, C_3, and v_e, and we can deduce the identifiability of the original parameters from that of the θ_i. Thus, we have for the second group

$$\frac{1 + \theta_4 s}{1 + s(\theta_1 + \theta_2 + \theta_3 + \theta_4) + s^2(\theta_1\theta_3 + \theta_1\theta_4 + \theta_2\theta_3 + \theta_3\theta_4) + s^3(\theta_1\theta_3\theta_4)} \quad (7.43)$$

From this the vector $\mathcal{M}_1(\theta)$ can be written as

$$\psi_1(\theta) = \begin{bmatrix} \theta_4 \\ \theta_1 + \theta_2 + \theta_3 + \theta_4 \\ \theta_1\theta_3 + \theta_1\theta_4 + \theta_2\theta_3 + \theta_3\theta_4 \\ \theta_1\theta_2\theta_3 \end{bmatrix} \quad (7.44)$$

To test for global identifiability it is necessary to show that for almost all $\hat{\theta}$.

$$\psi_1(\theta) = \psi_1(\hat{\theta}) \Rightarrow \theta = \hat{\theta} \quad (7.45)$$

This is equivalent to determining whether the following set of equations derived directly from Eq. (7.3.9) can lead to a unique solution of the parameters $\hat{\theta}$ in terms of θ.

$$\begin{aligned} \hat{\theta}_4 &= \theta_4 \\ \hat{\theta}_1 + \hat{\theta}_2 + \hat{\theta}_3 + \hat{\theta}_4 &= \theta_1 + \theta_2 + \theta_3 + \theta_4 \\ \hat{\theta}_1\hat{\theta}_3 + \hat{\theta}_1\hat{\theta}_4 + \hat{\theta}_2\hat{\theta}_3 + \hat{\theta}_3\hat{\theta}_4 &= \theta_1\theta_3 + \theta_1\theta_4 + \theta_2\theta_3 + \theta_3\theta_4 \\ \hat{\theta}_1\hat{\theta}_3\hat{\theta}_4 &= \theta_1\theta_3\theta_4 \end{aligned} \quad (7.46)$$

This is a system of polynomial equations that can be solved using the REDUCE routine for algebraic manipulation. It can also be solved by hand. By elimination of $\hat{\theta}_2, \hat{\theta}_3$, and $\hat{\theta}_4$ we obtain from Eq. (7.46)

$$\hat{\theta}_3^3 - \hat{\theta}_3^2(\theta_1 + \theta_2 + \theta_3 + \theta_4) + \hat{\theta}_3(\theta_1\theta_3 + \theta_1\theta_4 + \theta_2\theta_3 + \theta_3\theta_4) - \theta_1\theta_3\theta_4 = 0 \quad (7.47)$$

Since this equation has the trivial solution $\hat{\theta}_3 = \theta_3$, the left-hand side of Eq. (7.47) can be factored to obtain

$$(\hat{\theta}_3 - \theta_3)[\hat{\theta}_3^2 - \theta_3(\theta_1 + \theta_2 + \theta_4) + \theta_1\theta_4] \quad (7.48)$$

so that the model is not sgi and there are three possible local solutions for $\hat{\theta}_3$ as follows:

$$\hat{\theta}_3 = \theta_3 \tag{7.49}$$

$$\hat{\theta}_3 = (\theta_1 + \theta_2 + \theta_4 + \sqrt{\Delta_1})/2 \tag{7.50}$$

$$\hat{\theta}_3 = (\theta_1 + \theta_2 + \theta_4 - \sqrt{\Delta_1})/2 \tag{7.51}$$

with $\Delta_1 = (\theta_1 + \theta_2 + \theta_4)^2 - 4\theta_1\theta_4$.

It is now possible to calculate the values of $\hat{\theta}_1$ and $\hat{\theta}_2$ associated with each possible value of $\hat{\theta}_3$:

$$\hat{\theta}_1 = \theta_1\theta_3/\hat{\theta}_3 \tag{7.52}$$

$$\hat{\theta}_2 = \theta_1 + \theta_2 + \theta_3 - \hat{\theta}_3 - (\theta_1\theta_3/\hat{\theta}_3) \tag{7.53}$$

Thus, if we have one local solution for the model ψ_1, it should be possible to compute two additional sets of parameters. The three possible values of the original parameters are then readily calculated, using Eq. (7.42). Note that $\theta_4 = C_2/v_e$ is structurally globally identifiable so that there is only one value possible for θ_4, which is given by Eq. (7.46).

As an illustration of the method, the data corresponding to the ^{13}CO step-up run number 011680 of the paper by Otarod et al. (1983, Table 2) as in Example 7.3.2 were employed in conjunction with the previous equations to compute the three values for the four parameters corresponding to the model in Fig. 7.2. Values are listed in the following table, the first column being that obtained by computer as a local solution with parameter values and reported in the previous example.

Possible Values for $(\mathcal{M}_1)$ Model 1
Parameter Estimates*

Parameters	Set		
	1	2	3
$C_1, C^a l$ (ml/g)NTP	1.630	0.0168	0.262
$C_2, C^* l$ (ml/g)NTP	5.388	−8.785	5.850
$C_3, CH_x l$ (ml/g)NTP	0.164	15.90	1.020
v_e, (ml/min/g)NTP	0.177	−0.291	0.194

* $V = 0.33$ ml/g/min.

In this table set 2 can be ruled out on physical grounds because it lists negative values for the parameters. Of the other two local solutions, set 3 is preferable to set 1 originally generated in Example 7.3.2. Deuterium tracing data (Happel et al., 1982) indicate that the active carbon species concentration is relatively small compared with that of the hydrocarbon intermediate $CH_x l$.

Another matter to be considered is whether additional models could be constructed on the basis of a single rate-controlling mechanism that would fit the data equally well. This is the problem of distinguishability, addressed in the following example.

Example 7.3.4 Distinguishability in ^{13}CO Tracing for Methanation

The same reaction of methanation with the same ^{13}C data is employed to test the applicability of a different model, given by Fig. 7.3.

In this model the compartments designate the same species as previously, but it is assumed that the inactive carbon instead of exchanging with the active form can itself slowly react to produce the hydrocarbon intermediates.

The equations of state for this model are

$$\begin{aligned} C_1 \dot{x}_1 &= -Vx_1 + Vu \\ C_1 \dot{x}_2 &= -v_1 x_2 + v_1 x_1 \\ C_3 \dot{x}_3 &= -Vx_3 + (V - v_1)x_1 + v_1 x_2 \\ C_4 \dot{x}_4 &= -Vx_4 + Vx_3 \end{aligned} \quad (7.54)$$

Here the parameters to be estimated are C_1, C_2, C_3, and v_1. As in the case of identifiability it is first necessary to obtain the transfer function for this model, which we designate by the subscript 2.

Following the same procedure as previously, we have

$$\mathbf{H}_2(s, \theta) = \frac{1}{1 + (C_4/V)s} \quad (7.55)$$

$$\times \frac{1 + \left(\dfrac{C_2}{v_3} - \dfrac{C_2}{V}\right)s}{1 + s[(C_1/V) + (C_2/v_1) + (C_3/V)] + s^2[(C_1 C_2/Vv_1) + (C_1 C_3/V^2) + (C_2 C_3/v_1 V)] + s^3[(C_1 C_2 C_3/V^2 v_1)]}$$

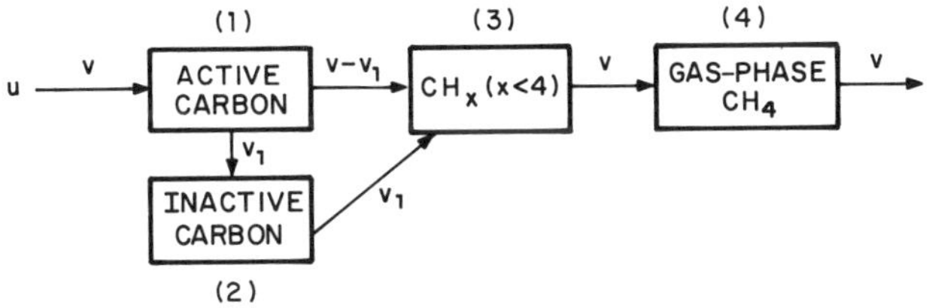

FIG. 7.3 Second compartmental model for ^{13}C tracing of methanation.

160   7 Transient Tracing

and defining the θ in the second factor by

$$\theta_1 = C_1/V, \quad \theta_2 = C_2/V$$
$$\theta_3 = C_3/V, \quad \theta_4 = C_2/v_1 \tag{7.56}$$

we write the second factor as

$$\frac{1 + (\theta_4 - \theta_2)s}{1 + s(\theta_1 + \theta_3 + \theta_4) + s^2(\theta_1\theta_4 + \theta_1\theta_3 + \theta_3\theta_4) + s^3(\theta_1\theta_3\theta_4)} \tag{7.57}$$

As before, from this the vector $\psi_2(\theta)$ can be written

$$\psi_2(\theta) = \begin{bmatrix} \theta_4 - \theta_2 \\ \theta_1 + \theta_3 + \theta_4 \\ \theta_1\theta_4 + \theta_1\theta_3 + \theta_3\theta_4 \\ \theta_1\theta_3\theta_4 \end{bmatrix} \tag{7.58}$$

We showed in Example 7.3.3 that model 1 was structurally locally identifiable and that a set of three possible solutions for this model made the most physical sense. Now in this example we wish to know whether, given any set of the parameters for model 1, it is possible to compute parameters that will satisfy the new model 2. If so, it would prove that model 2 is indistinguishable from model 1. This is equivalent to obtaining a solution for θ based on the following set of equations derived from Eqs. (7.44) and (7.58):

$$\theta_4 - \theta_2 = \hat{\theta}_4$$
$$\theta_1 + \theta_3 + \theta_4 = \hat{\theta}_1 + \hat{\theta}_2 + \hat{\theta}_3 + \hat{\theta}_4$$
$$\theta_1\theta_4 + \theta_1\theta_3 + \theta_3\theta_4 = \hat{\theta}_1\hat{\theta}_3 + \hat{\theta}_1\hat{\theta}_4 + \hat{\theta}_2\hat{\theta}_3 + \hat{\theta}_3\hat{\theta}_4 \tag{7.59}$$
$$\theta_1\theta_3\theta_4 = \hat{\theta}_1\hat{\theta}_3\hat{\theta}_4$$

Using the REDUCE algebraic manipulation routine, the following solution is obtained:

$$\theta_1 = \frac{\hat{\theta}_1 + \hat{\theta}_2 + \hat{\theta}_4 - \sqrt{(\hat{\theta}_1 + \hat{\theta}_2 + \hat{\theta}_4)^2 - 4\hat{\theta}_1\hat{\theta}_4}}{2}$$

$$\theta_2 = \hat{\theta}_3 - \hat{\theta}_4 \tag{7.60}$$

$$\theta_3 = \frac{\hat{\theta}_1 + \hat{\theta}_2 + \hat{\theta}_4 + \sqrt{(\hat{\theta}_1 + \hat{\theta}_2 + \hat{\theta}_4)^2 - 4\hat{\theta}_1\hat{\theta}_4}}{2}$$

$$\theta_4 = \hat{\theta}_3$$

This solution can be checked by substitution of values from Eq. (7.60) into Eq. (7.59).

Thus, model 2 cannot be distinguished from model 1, and by using values of the local parameters from model 1 ($\hat{\theta}$) we can compute a set of parameters that will also correlate the data.

In addition, we must now study the identifiability of model 2 in order to determine whether it can furnish more than the single set of parameters generated by Eq. (7.60) from those in model 1. This is accomplished by following the same procedure as in Example 7.33, but using the transfer function corresponding to Eq. (7.55) instead of Eq. (7.41). It is found that this model is not sgi and that there are six possible parameter sets that can be written using a hat "^" for the generating model:

$$
\begin{array}{lll}
(1)\ \theta_1=\hat{\theta}_1 & (2)\ \theta_1=\hat{\theta}_3 & (3)\ \theta_1=\hat{\theta}_4 \\
\ \ \ \ \theta_2=\hat{\theta}_2 & \ \ \ \ \theta_2=\hat{\theta}_2 & \ \ \ \ \theta_2=\hat{\theta}_2+\hat{\theta}_3-\hat{\theta}_4 \\
\ \ \ \ \theta_3=\hat{\theta}_3 & \ \ \ \ \theta_3=\hat{\theta}_1 & \ \ \ \ \theta_3=\hat{\theta}_1 \\
\ \ \ \ \theta_4=\hat{\theta}_4 & \ \ \ \ \theta_4=\hat{\theta}_4 & \ \ \ \ \theta_4=\hat{\theta}_3 \\
(4)\ \theta_1=\hat{\theta}_1 & (5)\ \theta_1=\hat{\theta}_3 & (6)\ \theta_1=\hat{\theta}_4 \\
\ \ \ \ \theta_2=\hat{\theta}_2+\hat{\theta}_3-\hat{\theta}_4 & \ \ \ \ \theta_2=\hat{\theta}_1+\hat{\theta}_2-\hat{\theta}_4 & \ \ \ \ \theta_2=\hat{\theta}_1+\hat{\theta}_2-\hat{\theta}_4 \\
\ \ \ \ \theta_3=\hat{\theta}_4 & \ \ \ \ \theta_3=\hat{\theta}_4 & \ \ \ \ \theta_3=\hat{\theta}_3 \\
\ \ \ \ \theta_4=\hat{\theta}_3 & \ \ \ \ \theta_4=\hat{\theta}_1 & \ \ \ \ \theta_4=\hat{\theta}_1
\end{array}
\qquad (7.61)
$$

This information is used first with the data for set 1 of model 1 as the basis for obtaining a solution for model 2. For this calculation the equations derived by REDUCE are employed, giving the first submodel listed in the following table. The remaining five submodels are computed using Eq. (7.61) with set 1 of model 1 as the generating model. The same six sets could have been obtained directly by starting with Eq. (7.58).

Possible Values for (M_2) Model 2 Parameter Estimates

Parameter	Set					
	1	2	3	4	5	6
C_1	1.020	15.900	0.164	1.020	15.900	0.164
C_2	−9.788	−9.788	5.948	5.948	−8.932	−8.932
C_3	15.900	1.020	1.020	0.164	0.164	15.900
v_1	−19.696	−19.696	0.1234	0.1234	−2.889	−2.889

Here, sets 1, 2, 5, and 6 can be rejected on the basis of negative values for the parameters. Set 4 is ruled out because of the extremely low value for $C_3 = \mathrm{CH}_x l$, as was the case for set 1 of the previous model. Thus, set 3 in the preceding table and set 3 in Example 7.3.2 (model 1) remain as possibilities.

To conclude, we cannot distinguish between model 1 and model 2. However, in spite of this, both models agree reasonably well in the values of surface concentrations of intermediates; $C_3 = CH_x l$ is 1.020 in both cases. The total carbon concentration C_1 and C_2 is 6.112, the same in both cases.

There have been many studies of the carbon intermediate in methanation. Results, as summarized by Biloen and Sachtler (1981), indicate the presence of a small concentration of active carbidic carbon intermediate and a larger layer of graphitic carbon that is less active and may indeed be a poison.

These examples serve to illustrate how conventional computer modeling may be augmented by considerations of identifiability and distinguishability. They also show how more qualitative information such as that obtainable by surface spectroscopies can be applied to resolution of identifiability problems arising in transient tracing.

There has been considerable continuing interest in the applications of transient isotopic tracing. Early studies, conducted largely in Hungary and Russia, are summarized in the book by Neiman and Gal (1971). A recent symposium devoted to the subject included diverse approaches to Fisher–Tropsch and related CO hydrogenation processing (Biloen *et al.*, 1985; Bell, 1985; Delgass *et al.*, 1985; McCandlish *et al.*, 1985; Walter *et al.*, 1985).

A study by Mirodatos (1986) employed C_6D_6 as a tracer in the hydrogenation of benzene to cyclohexane over a Ni/SiO_2 catalyst. Generally, approaches have been more qualitative than those discussed in this book.

Even though ^{18}O exchange cannot be modeled as quantitatively as ^{13}C exchange, it is also interesting for distinguishing catalytic mechanisms in which only a chemisorbed layer is involved from those comprising redox mechanisms in which sublayers of active catalyst components participate. Monnier and Keulks (1981) conducted a series of such studies on the catalytic oxidation of propylene. Using ^{18}O and deuterated propylene transients superposed on steady-state reactions conditions with a plug-flow reactor system using a β-$Bi_2Mo_2O_9$ catalyst, they concluded that oxidation to acrolein occurs via a redox mechanism with the involvement of sublayers of lattice oxygen.

In all of these studies the tracer species was employed in essentially unidirectional reactions. Those cases in which the tracer is employed in reversible reactions would be more difficult to model, although they are often identifiable in principle, as discussed in Chapter 6.

7.4 Tracing with Multiple Marked Species

When the molecule being traced contains more than a single atom of an element being marked, it is often possible to distinguish between the marked

7.4 Tracing with Multiple Marked Species

species. It may thus be possible to obtain more information than by taking only the fractional marking as the basis for dependent variables in the systems equations.

For example, Klisurski and Kuncheva (1980) have used oxygen isotope exchange to study the mechanism of oxidation reactions over oxide catalysts. The isotopic exchange reaction is

$$^{18}O_2 + {}^{16}O_2 \rightleftharpoons 2\,{}^{16}O^{18}O \qquad (7.62)$$

Redistribution of isotopic marked species was studied in the gas phase, both on previously marked and unmarked catalysts. Equations describing the transients are first order in time but they are nonlinear in concentrations of the three types of oxygen molecules.

For deuterium tracing of methanation the same problem of nonlinearity exists. However, by assuming that the known transient input is in equilibrium with the surface, it is possible to linearize the system equations. This technique is illustrated in the following example.

Example 7.4.1 Deuterium Tracing of Methanation

Deuterium marking was employed to study the methanation of carbon monoxide over a nickel catalyst using a gradientless recirculating reactor (Happel et al., 1982). The transient superposition technique was used. After steady state was reached with a given mixture of hydrogen and carbon monoxide, the feed stream was rapidly switched to one containing D_2/CO in exactly the same proportions as the H_2/CO content of the original feed. It was found by continuous monitoring of the effluent that there was only a small kinetic isotope effect with the commercial nickel on silica catalyst employed (Harshaw 104T). Thus, the transient tracer data correspond to steady-state operation of the catalyst in its working state.

The system is assumed to consist of a number of cells or compartments, each one containing one of the intermediates or terminal species. Reactants from one compartment are assumed to be equally accessible to those from another with which they interact. In examples discussed earlier in this chapter, the independent variables were taken as fractional marking of traced atoms in molecules traced.

With deuterium tracing such a simplification would result in the loss of considerable information conveyed by the distribution of deuterium among species containing more than a single deuterium atom. Therefore, the basic material balance equations were expressed in terms of the concentrations of deuterated species. It was still possible to retain a system of linear equations by expressing the fraction of hydrogen in the compartment containing chemisorbed hydrogen–deuterium as an observed function of time. It was assumed that the fraction of deuterium participating in the reaction from chemisorbed species was the same as that observed in the gas phase, corresponding to equilibrium. This assumption is supported by our own studies and those of other investigators. It provides the desired relationship for modeling of the hydrogenation of carbonaceous species and also enables them to be considered separately from steps involving water formation.

It was found that the fractional concentration of hydrogen in the gas phase for a given experiment could be accurately fitted by an exponential equation of the following form

$$f = C^{Hl}/C^o = \exp(-kt) \tag{7.63}$$

where f is the fraction of C^{Hl} on the surface of the total $Hl + Dl$ species; C^{Hl}, the concentration of Hl on the surface (ml/g) (NTP); C^o, the concentration of $Hl + Dl$ on the surface; it is assumed that $C^o = C^{Hl} + C^{Dl}$; k, a constant; and t, the time in minutes ($t = 0$ when D_2 is introduced).

Several models were employed for parameter determination. The most suitable one, which is illustrated here, is the following:

$$\begin{aligned}
H_2 + 2l &\rightleftharpoons 2H \quad \text{(equilibrium)} \\
CO + l &\rightleftharpoons COl \quad \text{(equilibrium)} \\
COl + l &\longrightarrow Cl + Ol \\
Cl + Hl &\longrightarrow CHl + l \\
CHl + Hl &\longrightarrow CH_2l + l \\
CH_2l + Hl &\longrightarrow CH_3l + l \\
CH_3l + Hl &\longrightarrow CH_4 + 2l
\end{aligned} \tag{7.64}$$

where l designates active surface sites on the catalyst.

With deuterium modeling there will be five terminal species, CH_4, CH_3D, CH_2D_2, CHD_3, and CD_4, and nine transient intermediates as the CH_x species are exchanged, CHl, CDl, CH_2l, $CHDl$, CD_2l, CH_3l, CH_2Dl, CHD_2l, and CD_3l.

The following 14 material balance equations can be written:

$$\frac{dC^{CDl}}{dt} = V\left(1 - f - \frac{C^{CDl}}{C_2}\right)$$

$$\frac{dC^{CHl}}{dt} = V\left(f - \frac{C^{CHl}}{C_2}\right)$$

$$\frac{dC^{CD_2l}}{dt} = V\left(\frac{C^{CDl}}{C_2}(1-f) - \frac{C^{CD_2l}}{C_3}\right)$$

$$\frac{dC^{CHDl}}{dt} = V\left(\frac{C^{CHl}}{C_2}(1-f) + \frac{C^{CDl}}{C_2}f - \frac{C^{CHDl}}{C_3}\right)$$

7.4 Tracing with Multiple Marked Species

$$\frac{dC^{CH_2l}}{dt} = V\left(\frac{C^{CHl}}{C_2}f - \frac{C^{CH_2l}}{C_3}\right)$$

$$\frac{dC^{CD_3l}}{dt} = V\left(\frac{C^{CD_2l}}{C_3}(1-f) - \frac{C^{CD_3l}}{C_4}\right)$$

$$\frac{dC^{CHD_2l}}{dt} = V\left(\frac{C^{CHDl}}{C_3}(1-f) + \frac{C^{CD_2l}}{C_3}f - \frac{C^{CHD_2l}}{C_4}\right)$$

$$\frac{dC^{CH_2Dl}}{dt} = V\left(\frac{C^{CH_2l}}{C_3}(1-f) + \frac{C^{CHDl}}{C_3}f - \frac{C^{CH_2Dl}}{C_4}\right) \quad (7.65)$$

$$\frac{dC^{CH_3l}}{dt} = V\left(\frac{C^{CH_2l}}{C_3}f - \frac{C^{CH_3l}}{C_4}\right)$$

$$\frac{dC^{CD_4}}{dt} = \frac{VW}{\beta}\left(\frac{C^{CD_3l}}{C_4}(1-f) - \frac{C^{CD_4}}{C_5}\right)$$

$$\frac{dC^{CHD_3}}{dt} = \frac{VW}{\beta}\left(\frac{C^{CHD_2l}}{C_4}(1-f) + \frac{C^{CD_3l}}{C_4}f - \frac{C^{CHD_3}}{C_5}\right)$$

$$\frac{dC^{CH_2D_2}}{dt} = \frac{VW}{\beta}\left(\frac{C^{CH_2Dl}}{C_4}(1-f) + \frac{C^{CHD_2l}}{C_4}f - \frac{C^{CH_2D_2}}{C_5}\right)$$

$$\frac{dC^{CH_3D}}{dt} = \frac{VW}{\beta}\left(\frac{C^{CH_3l}}{C_4}(1-f) + \frac{C^{CH_2Dl}}{C_4}f - \frac{C^{CH_3D}}{C_5}\right)$$

$$\frac{dC^{CH_4}}{dt} = \frac{VW}{\beta}\left(\frac{C^{CH_3l}}{C_4}(1-f) + \frac{C^{CH_4}}{C_5}\right)$$

where

$$C_2 = C^{CD} + C^{CH}$$
$$C_3 = C^{CD_2} + C^{CHD} + C^{CH_2}$$
$$C_4 = C^{CD_3} + C^{CD_2H} + C^{CDH_2} + C^{CH_3}$$
$$C_5 = C^{CD_4} + C^{CHD_3} + C^{CH_2D_2} + C^{CH_3D} + C^{CH_4}$$

The last term C_5 is an observed constant corresponding to the steady-state concentration in the product of the total CH_xD_{4-x} ($x = 0 - 4$); C^{il} is the concentration of ith species on catalyst (ml/g)(NTP) $i = CD, CH, CD_2, CHD, CH_2, CD_3, CHD_2, CH_2D$, and CH_3; C^i is the concentration of ith species in vapor phase, volume fraction; i, CD_4, CHD_3, CH_2D_2, CH_3D, and CH_4; $f = C^{Hl}/C_o = \exp(-kt)$; V is the velocity of CH_x production (ml/min/g$_{cat}$) (NTP); W is the weight of catalyst (g); and β is the dead space (ml).

This set of differential equations describes the population distribution of all species, intermediates and product components, according to the proposed mechanism. It can be expressed in matrix form as

$$d\mathbf{x}/dt = \mathbf{A}\mathbf{x} + \mathbf{b} \qquad (7.66)$$

where $\mathbf{x}$ is a vector of concentrations of all products and intermediates; $\mathbf{A}$, a matrix containing all the constants and the time-variant function f; and $\mathbf{b}$, a vector determined from the input of tracer in the feed that contains the time variant function f.

The 14 equations are not all independent because the sum of the concentration fractions for each of the adsorbed species must equal unity. Thus for example,

$$C^{CHl}/C_2 + C^{CDl}/C_2 = 1$$

There are four such relationships corresponding to the four parameters so that only ten of the equations are independent.

This model is structurally globally identifiable as shown in the previous chapter. In principle, a solution for *one* of the terminal species should provide sufficient information to determine C_2, C_3, and C_4. The parameter C_1 corresponding to the Cl concentration cannot be determined by deuterium tracing. In practice the curves giving the CH_4 and CD_4 transients can be fitted to a high degree of accuracy by expressions involving only one or two exponentials so they do not furnish sufficiently accurate information for discrimination between the parameters. Therefore, the intermediates that go through maxima in concentration are the most useful for providing information on values for assumed parameters which, of course, require expressions with several exponentials. Also, all ten of the independent differential equations are not required to fit the data for each separate observed deuteromethane. Thus, to fit the CH_2D_2 data requires eight equations, whereas the CH_3D and CD_3H concentration changes can be expressed by seven differential equations. In addition to requiring one additional equation for correlation, CH_2D_2 data showed the greatest scatter because of interference with other components. Therefore, the basic modeling procedure was carried out with the data for CH_3D and CHD_3 transients taken together. The surface concentrations thus obtained were used to plot values of the remaining species, CH_4, CH_2D_2, and CD_4.

A method that we had previously employed (Happel *et al.*, 1980) could, in principle, be used to solve the sets of first-order linear equations with time-dependent coefficients, but it proved difficult to obtain satisfactory convergence with the number of equations needed for deuterium tracing. We therefore took advantage of the fact that the reactions could be modeled as unidirectional by solving the equations analytically instead of using a computer routine to solve them numerically. Other features of the previously described method were retained including the statistical treatment that provides information on goodness-of-fit.

The following data were obtained in run number 021081

Inlet flow rate ml/min (NTP)		Outlet flow rates (during step-up) ml/min (NTP)		
H_2 or D_2	3.0	$H_xD_{2-x}(x = 0-2)$	1.01	Weight of catalyst = 2.43 g
CO	1.0	CO	0.24	Pressure = 1.0 atm
He	84.15	CO_2	0.05	Dead space = 118 ml
		$H_xD_{2-x}O(x = 0-2)$	0.66	Temperature = 210°C
		$CH_xD_{4-x}(x = 0-4)$	0.71	

These inlet rates were maintained constant during prereaction and step-up. The outlet rates were obtained during the step-up. At the beginning $x = 2$ and the species involved are H_2, H_2O, and CH_4. At the end $x = 0$ and we have D_2, D_2O, and CD_4 with CO and CO_2 being constant throughout the test.

Values observed and as-modeled for the terminal species are shown in Fig. 7.4. Agreement between computed and observed values is good. Results of parameter determination are summarized as follows:

$b_1 = CHl$, ml/g catalyst (NTP) $\quad = 4.44 \pm 0.06$

$b_2 = CH_2l$ $\quad = 0.54 \pm 0.01$

$b_3 = CH_3l$ $\quad = 0.21 \pm 0.00$

Average standard deviation of $CH_xD_{4-x}(x = 0 - 4) = 1.98$

The values of the parameters are well determined in spite of considerable random fluctuation in the analyses of the deuterated species. The correlation matrix is

$$\begin{bmatrix} 1.00 & 0.85 & 0.29 \\ 0.85 & 1.00 & 0.73 \\ 0.29 & 0.73 & 1.00 \end{bmatrix}$$

It indicates that the values for the three parameters are clearly defined, showing no correlation. This is reasonable because all intermediate step velocities are the same, as the reaction is far from equilibrium. If this were not the case, it would be necessary to consider step velocities as additional parameters that are not independent of surface concentrations.

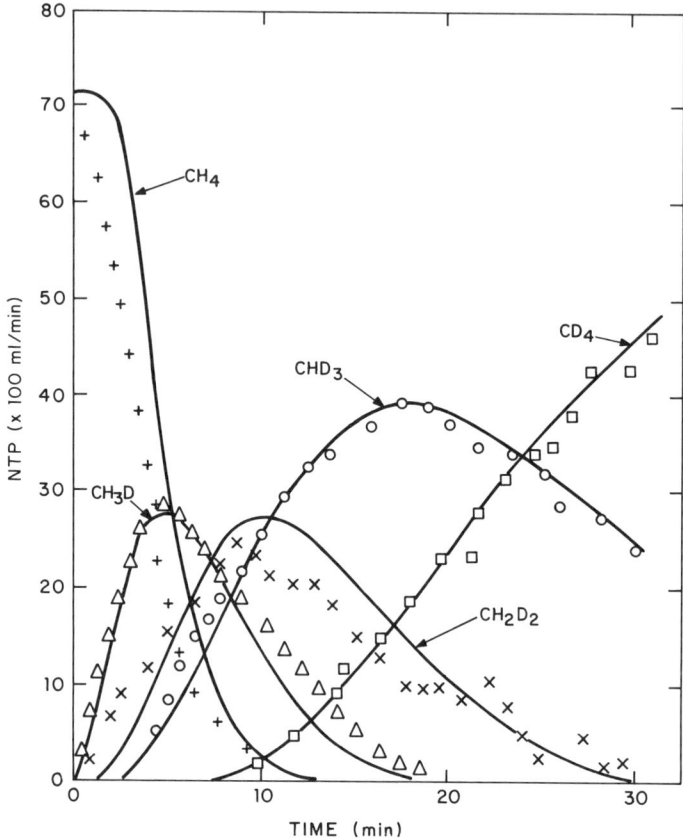

FIG. 7.4 Deuterium tracing, correlation of methanation.

Additional experiments were conducted in this study in which gaseous species were desorbed from the catalyst by helium after steady state was obtained. It was thought that, if the postulated model consisting of unidirectional hydrogenation of adsorbed species held exactly and if no reaction of surface species occurred during helium purging, it would be possible to titrate the adsorbed intermediates by injection of deuterium. In that case CHD_3 would be produced corresponding to CHl; CH_2D_2 would be produced corresponding to CH_2l; and CH_3D would be produced corresponding to CH_2l. This assumption was, in fact, borne out by fairly good agreement between values obtained by analysis of the effluent gases from such reaction. In addition CD_4 was produced corresponding to the presence of elemental carbidic carbon, C, chemisorbed on the catalyst. For run number 021081 reported in the example the corresponding Cl concentration on the catalyst was equivalent to 2.41 ml/g_{cat}. Separate hydrogen adsorption data on fresh catalyst indicate an adsorption at saturation of 7.1 ml/g of catalyst. Since hydrogen adsorbs dissociatively this amounts to 14.2 ml/g equivalent to a monolayer coverage of single active sites. It appears that the predominant hydrocarbon surface species is CHl, corresponding to more than all the other CH$_x l$ species together. Thus, the hydrogenation of CHl is a controlling step in the methanation process.

Mechanisms of this type in which a species (hydrogen in this case) is added at each step in the mechanism are of interest because they provide more information than catenary tracing, such as the conversion of ^{13}CO to methane. For unidirectional reactions, the mechanisms are structurally globally identifiable (sgi). Tracing hydrogen addition to nitrogen in forming ammonia affords additional examples of such a mechanism.

Another recent study (Happel et al., 1986) used deuterium and ^{13}CO tracing to study methanation over a molybdenum sulfide catalyst. An interesting feature was the finding that considerable structurally bound hydrogen is present on the working catalyst. It would be more difficult to conduct surface examination for adsorbed species on this type of catalyst as compared with metallic catalyst.

7.5 Conclusions

In this book, a major effort has been to separate the elucidation of reaction mechanisms from the development of kinetic equations in which both mechanistic and kinetic parameters are simultaneously involved. Tracing, and particularly transient tracing, afford a unique tool for accomplishing this. The determination of a mechanism by means of tracing experiments furnishes independent information, whereby the identification of critical steps can be accomplished. This, in turn, should be useful in the development or improvement of catalysts.

Once a tentative mechanism has been established, it is obvious that such critical steps will play an important role in the reaction kinetics. If we wish to develop a rate equation for practical application, independent knowledge of important mechanistic steps in advance should be useful. Rate equations based on such information, together with appropriate kinetic data, should possess a greater probability of extrapolation to a wider range of conditions than those developed on the basis of a more empirical computer modeling of the rate data alone.

As for assessment of reaction mechanisms, the previous chapters have represented an attempt to approach the subject from a systematic general viewpoint that includes not only tracing techniques. Mathematical development of methods for enumeration of possible reaction mechanisms and of identifiability and distinguishability concepts as applied to modeling should also extend beyond the subject of heterogeneous catalysis. The rationale developed can readily accomodate information now becoming available from a variety of developing surface spectroscopies.

The methanation reaction that we have treated at length in several examples is not the only system that might be advantageously studied. Some of the classic inorganic catalytic reactions such as the synthesis of ammonia and sulfuric acid were among the first to be studied by steady-state tracing. Transient tracing would add considerable insight into the mechanism of such reactions. A variety of other reactions could be readily studied since stable isotopes are available for practically all elements of interest. The use of ^{18}O as well as of ^{13}C and D for the study of organic reactions could be useful for the elucidation of the role of lattice oxygen in catalyst components and supports. Among organic syntheses those based on hydrogenation of carbon monoxide, hydrodesulfurization, oxidation, and ammo-oxidation of hydrocarbons should furnish immediate fruitful applications.

The final chapter considers the matter of how tracer data of the type obtained thus far can be applied to the subject of chemical kinetics. There is a large body of literature on this subject that will not be repeated. Our purpose is to call attention to ways in which tracer studies and principles associated with them may be helpful in rate expression development.

List of Symbols

Generally, vectors are represented by boldface lower case letters and matrices by boldface capital letters.

A	Matrix associated with states
B	NX by NP matrix defined by Eq. (7.2.9)
b	Input vector
Δ**b**	Parameter increment vector (see Eq. (7.2.21))
C_{kk}	kkth element of the variance–covariance matrix
C^i	Concentration of terminal species, fraction
C_i	Constants in Eq. (7.40)
C^{il}	Concentration of intermediate on catalyst surface
E_k	Error criterion for judging when convergence is achieved in a computer program operation
e	Parameter error vector for convergence
F^i_F	Rate of feed of component i to system
F^i	Rate of exit of component i from a system

f_j	Curve increment vector corresponding to number of parameters
g_m	Variable in variational equations corresponding to the number of system equations
$H_i(s,\theta), H_2(s,\theta)$	Transfer functions
I	Identity matrix
$\mathcal{M}_i$	Model
s	Laplace transform variable
t	Time
V	Overall reaction rate
v_e	Exchange velocity between active and inactive carbon
v_1	Reaction rate of inactive carbon
W	Weight of catalyst
x	State variable
z	State variable
β	Dead volume of system
ψ	Vector of parameter groupings in model $\mathcal{M}(\theta)$
θ	Parameter vector
$\Rightarrow$	implies that
$\hat{}$	hat to distinguish models
$\in$	belongs to

References

Amundson, N. R. (1966) "Mathematical Methods in Chemical Engineering." Prentice-Hall, New York.

Bard, Y. (1974). "Nonlinear Parameter Estimation." Academic Press, New York.

Bell, A. T. (1985). "Application of Transient Response Techniques to Quantify the Surface Concentrations of Species Participating in CO Hydrogenation," presented at 1985 Summer Nat. Meet. Am. Inst. of Chem. Eng. Seattle, Washington.

Berman, M. (1971). *Adv. Med. Phys. Symp. Pap. Int. Conf.*, 2nd, Boston, *1969*, pp 199–296.

Berman, M., Weiss, M. F. (1978), "SAAM 27 Manual." Englewood Cliffs DHEW Publication No. M. H. 78-180.

Biloen, P., and Sachtler, W. M. H. (1981). *Adv. Catal.* **30**, 165.

Biloen, P., and Zhang, X. (1985). "A Transient-Kinetic Study of Chaingrowth in the Fischer-Tropsch Synthesis," presented at 1985 Summer Nat. Meet. Am. Inst. of Chem. Eng. Seattle, Washington.

Box, G. E. P. (1960). *Ann. N. Y. Acad. Sci* **86**, 792.

Delgass, W. N., Miller, M. L., and Siddal, J. H. (1985). "Transient Kinetic Studies over Hydrocarbon Synthesis Catalysts," presented at 1985 Summer Nat. Meet. Am. Inst. of Chem. Eng. Seattle, Washington.

Draper, N. Y., Smith, H. (1968). "Applied Regression Analysis." Wiley, New York.

Fthenakis, V. (1978). "Studies on Methanation of Carbon Monoxide over Nickel Catalyst Using Transient Isotopic Tracing Technique." M. S. Thesis, Columbia University, New York.

Happel, J., Kiang, S., Spencer, J. L., Oki, S., and Hnatow, M. A. (1977). *J. Catal.* **50**, 429.
Happel, J., Suzuki, J., Kokayeff, P., and Fthenakis, V. (1980). *J. Catal.* **65**, 59.
Happel, J., Fthenakis, V., Suzuki, J., Yoshida, T., and Ozawa, S. (1981). *Proc. Int. Cong. Catal. 7th, Tokyo.*
Happel, J., Cheh, H. Y., Otarod, M., Ozawa, S., Severdia, A. J., Yoshida, T., and Fthenakis, V. (1982). *J. Catal.* **75**, 314.
Happel, J., Yoshikiyo, M., Yin, F., Otarod, M., Cheh, H. Y., Hnatow, M. A., Bajars, L., and Meyer, H. S. (1986). *I&EC Prod. Res. Devel.* **25**, 214.
Hearn, A. L. (1983). "Reduce Users' Manual." Rand Publication CP 78 (4/83), Santa Monica.
Kiang, S. (1978). "Transient Isotope Tracing Technique for the Resolution of the Room Temperature Oxidation of Carbon Monoxide Catalyzed by Commercial Hopcalite Catalyst." Doctoral dessertation, Columbia Univ., New York.
Klisurshi, D. G., and Kuncheva M. M. (1977). *Int. Chem. Eng.* **17**, 488.
Lanczos, Cornelius (1961). "Applied Analysis." Prentice-Hall, Englewood Cliffs, New Jersey.
Marquardt, D. W. (1963). *J. Soc. Ind. Appl. Math.* **11**, 4313.
McCandlish, L. E., Mims, C. A., and Melchior, M. T. (1985). "Isotopic Transient Studies of the Fischer–Tropsch Reaction," presented at 1985 Summer Nat. Meet. Am. Inst. of Chem. Eng., Seattle, Washington.
Monnier, J. R., and Keulks, G. W. (1981). *J. Catal.* **68**, 51.
Mirodatos, C. (1985). *J. Phys. Chem.* **90**, 481.
Otarod, M., Ozawa, S., Yin, F., Chew, M., Cheh, H. Y., and Happel, J. (1983). *J. Catal.* **84**, 156; *erratum* **89**, 564 (1984).
Siegel, S. (1956). "Nonparametric Statistics for the Behavioral Sciences." McGraw-Hill, New York.
Spencer, J. L., Long, C. L., and Kinny, J. M. (1971). *Ind. Eng. Chem. Fundam.* **10**, 2.
Spencer, J. L., Steiner, L., and Hartland, S. (1981). *AIChE J.* **27**, 1008.
Walter, E. (1982). Identifiability of State-Space Models, *Lect. Notes Biomath.* No. 46 Springer-Verlag, New York.
Walter, E., Lecourtier, Y., Bertrand, P., and Gomeni, R. (1980). *Proc. Joint Autom. Control Conf., San Francisco.*
Walter, E., Lecourtier, Y., Happel, J. and Kao, J.-Y. (1986). "Identifiability and Distinguishability of Fundamental Parameters in Catalytic Methanation." *AIChE J* **32**, 1360.

Chapter 8

Development of Rate Equations

8.1 Introduction

It has been shown that modeling with transient isotopic data enables direct determination of concentrations of intermediates as well as elementary step velocities. When kinetic equations alone are used for modeling, determination of these parameters is more indirect. Isotopic tracing is a technique that affords insight into reaction mechanisms so that before kinetic models are constructed significant additional information is available. To construct such rate equations requires further assumptions about how the elementary reaction step velocities are to be represented and how species compete with each other for occupancy of active catalyst sites. These subjects will be considered briefly in this chapter. An exhaustive treatment of the subject of kinetic modeling is beyond the scope of this book. The attempt will be made to indicate how such modeling can be improved in the context of the methods that have been presented in previous chapters.

As discussed in Chapter 3, the listing of direct mechanisms for a reaction system furnishes a useful procedure, once an initial choice of possible elementary steps has been made. Listing direct mechanisms provides a complete enumeration of all the possible cycle-free minimal mechanisms consistent with a given choice of steps. Mechanisms thought to be improbable on the basis of chemical theory or experimental evidence can then be discarded.

Sets determined in this way are different from what Horiuti (1973) and Temkin (1971) describe as routes, which they employ for the development of kinetic equations. The "routes" employed by these authors constitute linearly independent sets, which may be combined mathematically in ways resulting in the cancellation of elementary reaction steps. As Temkin (1979) notes, although the number of such basic routes is fixed for a given system, any set of routes can be employed as a basis, if the routes are linearly independent and their number is equal to the space of the routes. In this way Temkin arrives at mechanisms that may include cycles and derives rate equations from them. We believe that a preferable procedure is to employ cycle-free mechanisms for this purpose.

Once a suitable set of cycle-free direct mechanisms has been listed, appropriate rate equations may be developed following methods employed in standard chemical engineering texts on kinetics (Hougen and Watson, 1947; Carberry, 1976; Smith, 1981; Froment and Bischoff, 1979; Hill, 1977). These texts furnish comprehensive treatments of the subject of parameter estimation, largely in terms of rate equations based on Langmuir–Hinshelwood–Hougen–Watson (LHHW) models for simple reaction systems (those characterized by single overall reactions). It is assumed that the reader is familiar with these techniques. Usually the choice of mechanisms to be modeled is based on chemical theory and data but does not follow a systematic approach for combining elementary steps such as that discussed in Chapter 3. Rate equations corresponding to chosen mechanisms are tested on data obtained under varying conditions of temperature, total pressure, and gas-phase concentrations of reacting species. The preferred model is chosen on the basis of best fit to experimental data as judged by statistical criteria. Details of mathematical techniques for discrimination are also elaborated in such texts as that of Bard (1974).

Both steady-state (Chapter 5) and unsteady-state (Chapter 7) tracer data may often be used advantageously to reduce the number of models to be tested in such LHHW modeling. Steady-state experiments will often enable a determination of whether rates of transfer can be described in terms of a single rate-controlling path. Compartmental models used in transient tracing will reduce to two compartments, if a single rate-controlling step exists. In that case the rate-controlling step cannot be identified without knowledge of which steps are at equilibrium with each other. If conditions of reaction are far removed from equilibrium, it is reasonable to assume that no individual reaction step is at equilibrium and the rate-controlling steps can be identified by the presence of high reactant concentrations.

In the course of such modeling, it may also be desirable to consider problems of structural identifiability and distinguishability (Chapter 6) of the LHHW models directly in addition to the use of those concepts in compartmental modeling of tracer data. Such applications are discussed by Park and Himmelblau (1982) and Happel et al. (1986).

Systems treated in this book consist of steady-state reactions. In some heterogeneous catalytic systems it is possible for sustained oscillations in the generation of reaction products to occur (Slin'ko, 1980). For such conditions more complicated rate expressions will be required.

8.2 General Expression for the Kinetics of Simple Reactions

In developing LHHW models, parameter groupings will usually appear as coefficients of pertinent partial pressures of ambient species. Such expressions will be useful for engineering purposes although it is often not possible to establish the fundamental significance of the parameters involved. An important parameter in such rate expressions is the concentration of free sites Cl available for formation of chemisorbed species and intermediates. The usual assumption that the total population of all surface sites is fixed provides a relationship for establishing this parameter, but it is not possible to obtain empty site concentration by LHHW modeling or tracing to test the validity of this assumption.

An important point to be addressed is the question of when it is indeed possible to express a reaction rate in terms of a single expression. Relationships given in Sections 5.2 and 5.3 provide an answer to this question and afford some generalization of the LHHW models. It is, of course, only meaningful to discuss such models in the case of reversible reactions as developed in the texts mentioned previously. In such cases thermodynamic concepts can be applied to the derivation of the "potential factor" that occurs in these rate expressions.

For reversible reactions in which paths of transfer occur containing slow steps Eqs. (5.11) and (5.12) can be employed in the form

$$V^{1,2,3,\ldots,n} = V_+^{1,2,3,\ldots,n}[1 - (V_-^{1,2,3,\ldots,n}/V_+^{1,2,3,\ldots,n})] \tag{8.1}$$

This equation contains a velocity ratio that resembles that in Eq. (5.15) based on a theoretical development of step velocities in terms of thermodynamics. If instead of the stoichiometric number being different for each step in a sequence, a common stoichiometric number occurs that includes all the rate controlling steps in a mechanism, we have

$$V^{1,2,3,\ldots,n} = V_+^{1,2,3,\ldots,n}(1 - e^{\Delta G/v_r RT}) \tag{8.2}$$

Horiuti derived this equation for the case where v_r applied to only a single rate-controlling step in a sequence. This assumption is also made in the usual LHHW development. In that case, of course, the other steps will be at equilibrium with $v_{+s} = v_{-s}$ except for the rate-controlling step. Hougen and Watson's kinetic expressions reduce to a similar form. For an overall reaction of the type

$$aA + bB + \cdots \rightleftharpoons pP + qQ + \cdots \qquad (8.3)$$

Equation (8.2) can be expressed as

$$V = V_+ - V_- = V_+\left[1 - \left(\frac{p_P^p p_Q^q}{K_p p_A^a p_B^b}\right)^{(1/v_r)}\right] \qquad (8.4)$$

where K_p is the overall equilibrium constant for the reaction.

A unique way of determining whether a given reaction can be characterized by a factorable potential expression [shown in brackets in Eq. (8.4)] consists in measuring forward and reverse rates of transfer by steady-state tracing (Happel, 1978). For this purpose the forward and backward velocities should be studied not only close to equilibrium but over the entire range under which it is proposed to construct a rate equation. If reverse transfer of an atomic tracer is negligible, some steps will not be reversible and a rate expression involving a potential term will not be possible. For cases in which Eq. (8.4) applies, the development of an LHHW expression requires that only a single rate-controlling step exist.

This method of using tracers has been applied to several commercial reactions. In the case of isobutane dehydrogenation, it was found possible to extract a potential term and then to develop an appropriate rate equation using LHHW modeling (Happel et al., 1973). Data from some 340 experimental runs indicated that the best correlation was obtained by assuming a rate-controlling step consisting of nondissociative adsorption of isobutane.

The following example illustrates the derivation of the final equation employed and also serves to show how Eq. (8.4) can be employed to shorten the usual LHHW derivation procedure.

Example 8.2.1 Isobutane Dehydrogenation

The mechanism given in Example 5.4.1 is to be used as a model for the reaction with a single rate-controlling step, the chemisorption of isobutane with other mechanistic steps at equilibrium as usually assumed in LHHW modeling. The assumed mechanism is shown in the following table. Thus

$$V_+ = v_{+1} = k_{+1} p_{iC_4H_{10}} C_l$$

Since

$$1 = C_l + C_{H_2l} + C_{Hl} + C_{iC_4H_{10}l} + C_{iC_4H_8l} + C_{iC_4H_9l}$$

we need to evaluate the surface concentrations to obtain the free-site concentration C_l. It was found experimentally that no improvement in correlation was possible by including terms following the LHHW derivation involving more than the equilibrium adsorption constants for steps 4 and 6

8.2 General Expression for the Kinetics of Simple Reactions

Step number	Elementary reaction	Stoichiometric number v
1	$i\text{-}C_4H_{10} + l \underset{v_{-1}}{\overset{v_{+1}}{\rightleftarrows}} i\text{-}C_4H_{10}l$	1
2	$i\text{-}C_4H_{10}l + l \underset{v_{-2}}{\overset{v_{+2}}{\rightleftarrows}} i\text{-}C_4H_9l + Hl$	1
3	$i\text{-}C_4H_9l + l \underset{v_{-3}}{\overset{v_{+3}}{\rightleftarrows}} i\text{-}C_4H_8l + Hl$	1
4	$i\text{-}C_4H_8l \underset{v_{-4}}{\overset{v_{+4}}{\rightleftarrows}} i\text{-}C_4H_8 + l$	1
5	$2Hl \underset{v_{-5}}{\overset{v_{+5}}{\rightleftarrows}} H_2l + l$	1
6	$H_2l \underset{v_{-6}}{\overset{v_{+6}}{\rightleftarrows}} H_2 + l$	1

(Strangio, 1973):

$$K_4 = p_{iC_4H_8}C_l/C_{iC_4H_8l} \quad \text{and} \quad K_6 = p_{H_2}C_l/C_{H_2l}$$

so the free-site concentration is given by

$$C_l = \frac{1}{1 + (p_{iC_4H_8}/K_4) + (p_{H_2}/K_6)}$$

This is the so-called adsorption denominator that appears in LHHW kinetics. The model thus assumes the form

$$V = \frac{k_{+1}p_{iC_4H_{10}}}{1 + (p_{iC_4H_8}/k_4) + (p_{H_2}/k_6)}\left[1 - \frac{p_{iC_4H_8}p_{H_2}}{p_{iC_4H_{10}}K_{eq}}\right]$$

This expression gave the best correlation of experimental rate data. The following optimal values of the parameters k_1, K_4, and K_6 were obtained by a least-squares curve fitting procedure:

$$k_1 = 1.818 \times 10^3 \exp(-17.430/RT) \quad \text{moles/hr-g}_{cat}\text{-atm}$$

$$K_4 = 4.63 \times 10^4 \exp(-17,900/RT) \quad \text{atm}$$

$$K_6 = 2.33 \times 10^6 \exp(-24,100/RT) \quad \text{atm}$$

The overall equilibrium constant was determined experimentally since published values did not have the required accuracy

$$K_{eq} = 1.408 \times 10^7 \exp(-29,100/RT) \quad \text{atm}$$

Although the rate expression in Example 8.2.1 gave satisfactory correlation of data from a large number of models tested, the fundamental significance of the result of such LHHW modeling must still be questioned. In all cases the use of a squared denominator prefactor gave significantly poorer correlation than that using the first power. This implies that the rate-determining step is nondissociative chemisorption of isobutane. It seems more reasonable to assume that adsorption of a paraffin hydrocarbon like isobutane would be dissociative and lead to the formation of i-C_4H_9l. Another possibility would be the dual site conversion of i-C_4H_9l to i-C_4H_8l as a rate-controlling step.

Perhaps the result obtained is related to the basic LHHW assumption mentioned at the beginning of this section, namely that the active catalyst surface consists of a fixed total number of sites. The population of free sites is thus obtained as a difference that may be very small. Transient tracing might prove to be more suitable than steady-state tracing for establishing whether a single rate-controlling step exists for such a system because it can serve to identify a rate-controlling step more clearly.

Agreement with the LHHW theory is not as satisfactory for several other systems that were also studied by steady-state tracing. Thus, the development of rate expressions for SO_2 oxidation when traced by ^{18}O and ^{15}S showed that a potential term cannot be extracted to furnish a rate equation that is valid for the entire range of concentrations of interest. Some of the reasons for this are apparent from the following example.

Example 8.2.2 SO_2 Oxidation

It is desired to develop rate expressions for the catalytic oxidation of SO_2 according to the mechanism given in Examples 5.3.1 and 5.5.1. The assumed reaction mechanism is

Step number	Reaction step	Stoichiometric number v
1	$O_2 + 2l \underset{v_{-1}}{\overset{v_{+1}}{\rightleftharpoons}} 2Ol$	1
2	$SO_2 + l \underset{v_{-2}}{\overset{v_{+2}}{\rightleftharpoons}} SO_2l$	2
3	$SO_2l + Ol \underset{v_{-3}}{\overset{v_{+3}}{\rightleftharpoons}} SO_3l + l$	2
4	$SO_3l \underset{v_{-4}}{\overset{v_{+4}}{\rightleftharpoons}} SO_3 + l$	2
	$2SO_2 + O_2 = 2SO_3$	

From Example 5.4.1 the general thermodynamic relationship is given as

$$\frac{K_p p_{SO_2}^2 p_{O_2}}{p_{SO_2}^2} = \frac{v_{+1}}{v_{-1}} \left(\frac{v_{+2} v_{+3} v_{+4}}{v_{-2} v_{-3} v_{-4}} \right)^2$$

From this, there are two possible ways to obtain rate expressions with separable potential terms by using Eqs. (5.6) and (5.7) for either of the follow-

8.2 General Expression for the Kinetics of Simple Reactions

ing two paths of atomic transfer:

(1) The path over which atomic oxygen can proceed from reactant O_2 through steps 2, 3, and 4 to the product SO_3.

(2) The path over which either atomic oxygen or atomic sulfur can proceed from reactant SO_2 through steps 1, 3, and 4 to the product SO_3.

One type of equation with a potential term would be possible if steps 2, 3, and 4 were at equilibrium with step 1 rate controlling. From 8.4, this would be

$$V = v_{+1}\left(1 - \frac{p_{SO_3}^2}{K_p p_{O_2} p_{SO_2}}\right)$$

On the other hand, if step 1 were at equilibrium, we could write

$$V = V_+^{2,3,4}\left[1 - \left(\frac{p_{SO_3}^2}{K_p p_{O_2} p_{SO_2}}\right)^{1/2}\right]$$

where the velocity $V_+^{2,3,4}$ would be obtained from Eq. (5.6).

Thus, for oxygen chemisorption (step 1) control

$$v_{+1} = k_{+1} p_{O_2} C_l^2$$

Since $1 = C_l + C_{SO_2 l} + C_{SO_3 l} + C_{Ol}$, we need to evaluate the appropriate surface concentrations:

$$v_{+2} = v_{-2}, \quad \text{so} \quad k_{+2} p_{SO_2} C_l = k_{-2} C_{SO_2 l}$$

$$C_{SO_2 l} = \frac{k_{+2}}{k_{-2}} p_{SO_2} C_l = K_2 p_{SO_2} C_l$$

$$v_{+4} = v_{-4}, \quad \text{so} \quad k_{+4} C_{SO_3 l} = k_{-4} p_{SO_3} C_l$$

$$C_{SO_3 l} = \frac{1}{K_4} p_{SO_3} C_l$$

$$v_{+3} = v_{-3}, \quad \text{so} \quad k_{+3} C_{SO_2 l} C_{Ol} = k_{-3} C_{SO_3 l} C_l$$

$$C_{Ol} = \frac{C_{SO_3 l}}{K_3 C_{SO_2 l}} C_l = \frac{p_{SO_3}}{K_2 K_3 K_4 p_{SO_2}} C_l$$

$$1 = C_l + K_2 p_{SO_2} C_l + \frac{1}{K_4} p_{SO_3} C_l + \frac{1}{K_2 K_3 K_4} \frac{p_{SO_3}}{p_{SO_2}} C_l$$

$$C_l = \frac{1}{1 + K_2 p_{SO_2} + (1/K_4) p_{SO_3} + (1/K_2 K_3 K_4)(p_{SO_3}/p_{SO_2})}$$

hence

$$V = \frac{k_{+1} p_{O_2}[1 - (p_{SO_3}^2/K_p p_{O_2} p_{SO_2})]}{[1 + k_2 p_{SO_2} + (1/K_4) p_{SO_3} + (1/K_2 K_3 K_4)(p_{SO_3}/p_{SO_2})]^2}$$

For the case with steps 2, 3, and 4 not at equilibrium, it is not possible to separate C_l without assuming a single rate-controlling step or using an empirical relationship for $V_+^{2,3,4}$. If, for example, we assume that SO_3 desorption is rate controlling $V_+^{2,3,4} = v_{+4} = k_{+4} C_{SO_3 l}$.

$v_{+2} = v_{-2}$, so $C_{SO_2 l} = K_2 p_{SO_2} C_l$

$v_{+1} = v_{-1}$, so $K_{+1} p_{O_2} C_l^2 = k_{-1} C_{Ol}^2$, $C_{Ol} = K_1^{1/2} p_{O_2}^{1/2} C_l$

$v_{+3} = v_{-3}$, so $C_{SO_3 l} = K_3 (C_{SO_2 l} C_{Ol}/C_l)$

$\qquad = K_3 K_2 p_{SO_2} C_l (K^{1/2} p_{O_2}^{1/2} C_l / C_l)$

$\qquad = K_2 K_3 K_1^{1/2} p_{SO_2} p_{O_2}^{1/2} C_l$

$1 = C_l + K_2 p_{SO_2} C_l + K_1^{1/2} K_2 K_3 p_{SO_2} p_{O_2}^{1/2} C_l + K_1^{1/2} p_{O_2}^{1/2} C_l$

$C_l = 1/(1 + K_2 p_{SO_2} + K_1^{1/2} K_2 K_3 p_{SO_2} p_{O_2}^{1/2} + K_1^{1/2} p_{O_2}^{1/2})$

Hence

$$V = \frac{k_{+4} K_1^{1/2} K_2 K_3 p_{SO_2} p_{O_2}^{1/2} [1 - (p_{SO_3}^2/K_p p_{O_2} p_{SO_2})^{1/2}]}{(1 + K_2 p_{SO_2} + K_1^{1/2} K_2 K_3 p_{SO_2}^{1/2} p_{O_2}^{1/2} + K_1^{1/2} p_{O_2}^{1/2})}$$

These two rate equations with single rate-controlling steps could also have been obtained using the formal LHHW procedure, in which the forward and reverse velocity of the rate-controlling step is first evaluated, as described previously. The general development is discussed by Happel and Csuha (1971a,b).

Data could not be satisfactorily correlated by either of the two relationships given in this example, nor could a good correlation be obtained by a number of other equations proposed by various investigators. It appears that both oxygen chemisorption and sulfur trioxide desorption may control the rate of oxidation of sulfur dioxide, as discussed by Happel and Mezaki (1974). Thus, it may not be possible to obtain an exact rate expression for this reaction. An empirical correlation was attempted by resorting to expressions in which the exponent of the term containing the equilibrium constant lies between $\frac{1}{2}$ and 1.

In view of the complicated nature of the LHHW expressions, it has been suggested by Weller (1956) that rate expressions in the empirical power law form be used to express forward reaction rates:

$$V_+ = k p_A^m p_B^n p_R^q \cdots \qquad (8.5)$$

instead of attempting to use separate kinetic and adsorption denominator expressions. The difficulty with this procedure is that, if chemisorbed products

8.2 General Expression for the Kinetics of Simple Reactions

are expressed by partial pressures taken to negative exponents, the expression for forward rate approaches infinity at zero concentration of such products. Expressions derived by the LHHW technique are also not entirely free from this problem. Thus, if we chose to use the rate expression in Example 8.2.2 corresponding to oxygen chemisorption as rate controlling, the numerator of the reverse velocity expression would be $k_{+1} p_{SO_3}^2 / K_p p_{SO_2}$ so that the reverse rate of SO_3 decomposition would appear to become extremely large at low SO_2 concentrations.

What is required is a rate expression that enables a finite predicted reaction velocity to be obtained for a reversible reaction over the entire range of concentrations of components. If equilibrium conversion is high the appropriate expression cannot include denominator factors with positive partial pressures of reactants. Modeling by the LHHW method can furnish guidance for the powers to be employed for species far from equilibrium and for the form of the adsorption denominator. Once this form has been obtained either by LHHW modeling or by empirical power law rate expressions, it must be modified to provide a form that will not lead to improbable rate expressions near equilibrium.

This can be accomplished by making use of the approximation that a power law expression can be converted to an equivalent fractional form such that the numerator will have powers equal to the coefficients of the overall reaction being modeled. The following example illustrates the procedure.

Example 8.2.3

A rate equation due to Calderbank (1952) gives the empirical relationship

$$V_+ = k p_{O_2}^{0.8} p_{SO_2}^{0.4}$$

for the catalytic oxidation of SO_2. It is desired to obtain a thermodynamically consistent rate expression as equilibrium is approached, assuming that close to equilibrium, desorption of SO_3 will be rate controlling. From Example 8.2.2 we have

$$V = V_+ \left[1 - \frac{p_{SO_3}}{K_p^{1/2} p_{O_2}^{1/2} p_{SO_2}^{1/2}} \right]$$

The expression for V_+ is converted to the form:

$$V_+ = \frac{k_1 p_{O_2}^{0.5} p_{SO_2}^{0.5}}{1 + k_2 p_{O_2} - k_3 p_{SO_2}}$$

This will lead to the reverse velocity

$$V_- = \frac{k_1 p_{SO_3} / K_p^{0.5}}{1 + k_2 p_{O_2} - k_3 p_{SO_2}}$$

Such an expression will approach equilibrium without unusual kinetics for the reverse reaction. Without modification of the Calderbank expression, we would obtain

$$V_- = \frac{k p_{SO_3} p_{O_2}^{0.3}}{K_p^{1/2} p_{SO_2}^{0.1}}$$

On the other hand the LHHW expression for the reverse velocity in the case of SO_3 desorption controlling from Example 8.2.2 would be

$$V_- = \frac{k_{+4} K_1^{1/2} K_2 K_3 p_{SO_2}^{1/2} p_{SO_3}}{K_p^{1/2}(1 + K_2 p_{SO_2} + K_1^{1/2} K_2 K_3 p_{SO_2} p_{O_2}^{1/2} + K_1^{1/2} p_{O_2}^{1/2})}$$

The last two expressions exhibit unreasonable behavior at high degrees of conversion, corresponding to very low concentrations of O_2 or SO_2. If the rate-controlling step in a reaction mechanism is irreversible, the thermodynamic considerations discussed in the previous paragraphs no longer apply. Such irreversibility is readily tested by marking product species with isotopes and determining whether tracer is detected in the initial reactants. With the assumption of only a single irreversible, and therefore rate-controlling step, the derivation proceeds following the usual LHHW technique. A rate equation will be obtained in which the potential term is absent. If a number of steps is irreversible, it will be possible to determine the one that corresponds to the largest concentration of a surface species by transient tracing. To employ the LHHW technique this step is chosen as rate controlling with other steps taken at equilibrium.

The extension of LHHW models to cases in which more than a single rate-controlling step exists, results in complicated rate expressions. In the case of reversible reactions it is still often useful to employ steady-state tracer data in conjunction with Eq. (5.15) to determine whether in priniciple a rate expression with a separate potential term exists in such cases. This application is discussed by Happel and Mezaki (1970) and Mezaki *et al.* (1980) in connection with ethanol and secondary butanol dehydrogenations.

There have been a number of studies on the methanation of carbon monoxide over nickel catalyst that serve to illustrate extensions of the LHHW approach to models with more than a single rate-controlling step and involving irreversible steps. The following example gives typical results.

Example 8.2.4 Methanation of Carbon Monoxide with Nickel Catalyst

Van Meerten *et al.* (1982) reviewed a number of proposed methods for correlation of methanation data and presented a new rate equation including the assumption of hydrogenation of CH*l* as a rate-determining step. This is consistent with studies by Happel *et al.* (1980, 1982) and Otarod *et al.* (1983) based on transient tracing experiments with D and ^{13}C isotopes. They also assumed the final step in hydrogenation of oxygen to be unidirectional,

$$OHl + Hl \longrightarrow H_2O + 2l$$

and presented an LHHW-type derivation to accommodate the assumption of two unidirectional steps. A considerable amount of experimental data spanning five orders of magnitude in methanation rate enabled the complicated rate equations obtained to be simplified to the form

$$V = \frac{Z_1 p_{CO}^{1/2}}{[1 + Z_2(p_{CO}/p_{H_2})^{1/2}]^2}$$

with only the two parameters Z_1 and Z_2. This equation gave an excellent fit to the data.

Other schemes similar to LHHW models have also been developed using so-called Rideal–Ely mechanisms in which direct reaction of gas-phase species with adsorbed intermediates occur. These models have not found as wide acceptance in mechanistic studies, perhaps because it is difficult to distinguish such a situation from one in which very low concentrations of adsorbed species are present.

Attempts have also been made to employ more complicated models for chemisorption than the Langmuir–Hinshelwood equation, particularly by Temkin (1979). He points out that insofar as adsorption equilibrium is concerned better agreement can be obtained by discarding at least one of the initial assumptions. Thus, either adsorption sites might be taken as differing from each other energetically or a mutal influence of adsorbed sites might be assumed. Temkin has developed in some detail kinetic models based on the former assumption. The modeling of heterogeneous catalytic systems has been treated in a monograph in Russian by Snagovskii and Ostrovskii (1976). Much of the Russian literature is reviewed including Temkin's important contributions.

8.3 Kinetics of Multiple Reactions

In Chapter 3 a system was developed for expressing the mechanisms of multiple overall reactions that were characterized by the requirement of more than a single chemical equation to express the observed rates of change of terminal species. It was shown that each mechanism could be written in terms of arbitrary degrees of advancement for the elementary steps and that these rates were related to the overall reactions involved. The following example using the system treated in Example 3.6.2 for ethylene oxide production illustrates how mechanisms are related to overall reactions among terminal species.

Example 8.3.1 The Space of Overall Reactions in Ethylene Oxide Production

In the mechanisms developed for ethylene oxide synthesis, it was shown that reactions among the five terminal species O_2, C_2H_2, C_2H_4O, CO_2, and H_2O could be expressed for certain mechanisms in terms of two specific reactions operating independently. The form in which the overall reaction is written suggests that it is made up of the reactions:

$$O_2 + 2C_2H_4 = 2C_2H_4O$$
$$3O_2 + C_2H_4 = 2CO_2 + 2H_2O$$

However, this is not inherent in the chemical system. Actually there are four simple reactions in the space of overall reactions, their equations being

$$O_2 + 2C_2H_4 = 2C_2H_4O$$
$$3O_2 + C_2H_4 = 2CO_2 + 2H_2O$$
$$5O_2 + 2C_2H_4O = 4CO_2 + 4H_2O$$
$$2CO_2 + 2H_2O + 5C_2H_4 = 6C_2H_4O$$

This set is obtained by a procedure that is analogous to the procedure for finding direct mechanisms. We systematically eliminate columns of the reaction-by-species matrix. If the matrix has rank R, we look at $R - 1$ independent columns at a time. In this example $R = 2$, so we eliminate one column at a time. Each addition of elements in the remaining columns produces one of the simple reactions. In this case two of these are replicated so the above four reactions are obtained.

The range of concentrations of the five terminal species over which rate equations could be written for the six possible combinations would depend on thermodynamic constraints and chemical reaction rates. Transient tracing might furnish additional pertinent restrictions.

As in the case of simple reactions, it is also possible to consider separately the effects of the intrinsic kinetics of the set of elementary steps in a given mechanism and the effects due to competitive adsorption of terminal species and intermediates on active catalyst sites, using either an LHHW or empirical development. The forms of LHHW equations become complicated and sufficient data are not yet available to assess their usefulness.

The following example shows a typical application using a more empirical approach than that used in Example 8.2.4. That example was devoted to the methanation of carbon monoxide over a Ni/SiO_2 catalyst, which could be expressed by a simple reaction. In this example, methanation is conducted over a transition metal sulfide catalyst, Happel *et al.* (1983). The kinetics involves two simple reactions rather than just a single one.

8.3 Kinetics of Multiple Reactions

Example 8.3.2 Methanation of Carbon Monoxide—Transition Metal Catalysts

Transition element sulfides catalyze the direct methanation of carbon monoxide without the need to first produce hydrogen by the water-gas shift reaction. Happel *et al.* (1983) studied the kinetics of this system on a bench-scale unit at pressures ranging from 1.48 to 6.99 MPa (19.6–69.0 atm) temperatures from light-off to 649°C and various space velocities with integral conversion. Under initial reaction conditions with feeds consisting predominantly of H_2 and CO it was found that practically no water was produced. However, at high conversions the CO_2 produced reacted, so that under these conditions H_2O was also formed to a small extent. In the analysis of the possible mechanisms, Happel and Sellers (1983) indicated that a multiple reaction consisting of two simple overall reactions could describe transformations among the five species involved: H_2, CO, CO_2, H_2O, and CH_4. Any two of five simple reactions could be chosen over a limited range of concentrations. However, at initial conditions of reaction, only two reactions could be employed. To obtain two rate expressions that would be applicable over the range of interest, the following were chosen since production of water does not occur until considerable CO_2 has been formed:

$$2H_2 + 2CO = CH_4 + CO_2$$

$$H_2 + CO_2 = CO + H_2O$$

Data were correlated for a number of runs using computer modeling of the two simultaneous reactions producing methane and water, respectively, as hydrogenation products. The following rate equations were found empirically to apply to performance at 520°C:

$$r_{CH_4} = \frac{22.67 p_{CO} p_{H_2}^{0.5}}{(1.0 + 0.085 p_{CO_2})^{1.15}}$$

$$r_{H_2O} = 0.296 p_{CO_2} p_{H_2} \left(1.0 - \frac{p_{CO} p_{H_2O}}{0.2208 p_{CO_2} p_{H_2}} \right)$$

where r_{CH_4} is the rate of methane production per hour per unit volume of reactor $(m^3(NTP)/hr/m^3)$; r_{H_2O}, the rate of water production per hour per unit volume of reactor $(m^3(NTP)/hr/m^3)$; and p^i, the partial pressure of species i (i = CO, H_2, CO_2, H_2O) at reaction conditions (atm) (1 atm = 0.1013 MPa).

The production of water is expressed as the reverse of the water-gas shift reaction; $K_p = 0.2208$ is the correct equilibrium constant for this reaction. It was not accurate to employ the true equilibrium constant, because of the simplified expression used for r_{H_2O}. The term in parentheses containing K_p is the so-called potential term, which approaches zero when the reaction is close to equilibrium. This term is necessary to correlate the data in the case of feeds containing carbon dioxide. The reverse water-gas shift reaction occurs at first but as the reactions proceed the consumption of hydrogen by methanation eventually leads to the consumption of water initially produced. It is therefore necessary to express the reverse water-gas shift reaction in a form that is valid in both directions. The equation employed accomplishes this as discussed in Example 8.2.3. For conditions corresponding to those modeled, the methanation reaction itself is not close to equilibrium and always corresponds to the production of methane so it is not necessary to employ a potential term.

In systems where irreversible steps occur as in methanation (Example 8.2.4) it is not possible to rigorously support the assumption of a single rate-controlling step. The satisfactory use of LHHW kinetics in Example 8.2.1 involved the reversible dehydrogenation of isobutane. It might be thought that for multiple reactions that are reversible more accurate modeling would also be possible. Studies by Happel et al. (1970) on the dehydrogenation of n-butane to produce an isomeric mixture of n-butenes indicated that this is not the case.

To summarize let us consider the convenient generalized rate representation of LHHW kinetics proposed by Yang and Hougen (1950) for a simple reaction

$$V = \text{(kinetic term)(potential term)/(adsorption term)} \tag{8.6}$$

This equation can also be applied to multiple reaction systems with little modification in principle. The potential term for each reaction can be written on the basis of known thermodynamic data. The kinetic term depends only on the law of mass action and is readily stated provided we can identify a rate-controlling step for each simple reaction. However, the adsorption denominator collects most of the hard-to-justify additional assumptions of the technique—uniform, noninteracting sites of only one kind. Thus, it appears that provisionally we might use LHHW modeling with the assumption of a single rate-controlling step for each simple reaction. We then obtain expressions for the simple reactions in which the unoccupied surface site concentration C_l is undetermined. It is this factor that comprises the adsorption denominator that must be, at present, derived on the basis of additional experimental data. If, in the case of a multiple reaction, there is no competition between active sites for each of the simple component reactions the adsorption denominators will be different. The rate of each simple reaction may be influenced by species common to both reactions.

Additional theoretical and experimental development will be needed to make further progress in the construction of rate expressions that can be extrapolated with confidence. It is believed that the employment of transient data will be useful for this purpose, since it provides a unique method for determination of step velocities and surface concentrations of intermediates corresponding to a given set of reaction conditions.

8.4 Concluding Remarks

The object in this book has been to explore the use of isotopic tracer techniques for furnishing additional insight into the mechanistic details of heterogeneous catalytic reaction systems. These systems, as in the case of most chemical reactions, consist of a number of elementary steps. In addition, they possess the two additional characteristic features of being catalytic and involving more than a single phase. The interaction of an ambient phase, usually gaseous, with a solid catalyst constitutes the source of major problems in correlation. The passage of tracers discussed here has been modeled as a rate process, either steady or transient, and tracers have been employed to follow the movement of atomic species between reactants and products, in both forward and reverse directions. This enables us to concentrate on the dynamics of a reaction system.

The transport of chemical species from the gas phase to the chemisorbed state and the interactions of chemisorbed species are more complicated than processes taking place in a homogeneous phase. In many chemical engineering applications it is possible to assume that transport phenomena between phases can be visualized as a combination of diffusional resistances with the species at the interface itself being at equilibrium with each other. In the case of physical adsorption it is generally assumed that the interaction between phases is sufficiently rapid so that diffusion through idealized films constitutes the major resistance to transport. For chemisorption this is not the case and equilibrium may not be reached even in the absence of diffusional resistance. The Langmuir adsorption theory is a first approximation to a description of chemisorption, but it is well known that it does not accurately describe chemisorption and desorption even in the absence of additional reaction steps.

Before accurate representations of chemisorption can be made, it will be necessary to know more exactly how adsorbed species interact with the solids on which they chemisorb. The study of chemisorbed species *in situ* has been an area in which considerable progress has been made in recent years under the general subject of surface science. A variety of sophisticated spectroscopic techniques has been developed. One of the most useful is X-ray photoelectron spectroscopy (XPS), and excellent studies of chemisorbed species on metal crystal surfaces have been reported using this method. The technique requires operation at very low pressures (about 10^{-8} torr), and thus it is not simple to conduct measurements while a reaction is actually taking place. Infrared (IR) spectroscopy does not possess this disadvantage but it has been difficult to operate with sufficient sensitivity to detect surface species during reaction, especially under transient conditions desirable for distinguishing between reaction intermediates and nonreacting adsorbed species.

Without entirely satisfactory methods for modeling chemisorption, it is difficult to develop kinetic equations for modeling reaction rates on a theoretical basis. It is still possible, however, to obtain useful information about the mechanism of heterogeneous catalytic reactions from a combination of information available via surface spectroscopies and kinetic studies.

The use of isotopic tracers employing both steady state and transient techniques has been advanced here as a valuable kinetic technique that enables much more mechanistic information to be obtained than the usual overall reaction kinetics. This technique can often be profitably supplemented by other kinetic techniques such as desorption of adsorbed intermediates produced during steady-state operation. Desorption following a rising temperature program (TPD) and reactive removal of adsorbed species (e.g., titration of carbonaceous intermediates by deuterium) present useful additional variations. Methods of surface science along with basic physical chemical principles such as transition-state theory and thermodynamics are necessary to postulate the set of possible mechanistic steps for a given system. We have attempted to show how it is possible to proceed systematically beyond this point to the construction of reaction models from which desired mechanistic parameters can be deduced from tracer experiments. The parameters obtainable from such modeling are velocities of reaction steps and concentrations of surface intermediates of mechanisms that can generally be shown to have more than purely empirical significance.

Such information can be of immediate value in catalyst development. In testing several catalysts of different activity it is possible to pinpoint improved activity in terms of specific slow steps involved rather than only in terms of overall activity. If there are several irreversible steps in the mechanism for a given reaction, they will all have the same velocity. However, a rate-controlling step can be identified as one associated with the highest surface concentration of intermediate since this would usually be the step with the lowest reaction velocity constant. The effect of change in catalyst formulations on a rate-controlling step should be a useful index in identifying the importance of various changes in composition.

Compartmental modeling techniques discussed in this book generally will only enable information to be obtained concerning the portion of the intermediates associated with the elements comprising the terminal reactants and products. Information on how the intermediates are associated with catalyst components is sometimes available by surface spectroscopy. In cases where one of the elements present in the catalyst, such as sulfur or oxygen, is also present in terminal species, it may be possible to determine whether the catalyst serves only as a template to accumulate intermediates in effective configuration or whether its component elements enter into catalyst composition.

When it is desired to proceed further to develop kinetic equations for reactions over a range of composition, temperature, and pressure, some type of basic correlating relationships must be used beyond the material balances employed in compartmental modeling. Compartmental modeling may give the step velocity and intermediate concentrations for a given set of operating conditions. If sufficient data are obtained, it should be possible to tabulate such rates for direct use in computer programs. What is wanted in addition is kinetic relationships that can be extrapolated beyond conditions used in experiments for catalyst evaluation.

The most useful general procedure thus far developed for this purpose is the LHHW technique. While this method has shortcomings, it is still widely used. The *a priori* knowledge available by tracing of whether a potential term can be extracted to form a rate expression and whether a rate-controlling step can be identified should be useful in implementing the development of LHHW expressions. Limited experience indicates that the adsorption denominator terms in such expressions should be modified by empirical procedures.

As for the future, further progress in modeling will be substantially advanced when surface science has developed to the point where isotopic transients of surface species can be measured during reaction with the same facility as is now possible for terminal species by mass spectrometry. Modeling equations automatically include the intermediate transients as additional dependent variables so that, if they were available as data, identification would be much sharper.

For the present, steady-state and transient isotopic tracing affords a technique that can be used alone or in conjunction with other techniques and theories to obtain useful mechanistic and kinetic information on heterogeneous catalysis systems. It can be applied to experimental or industrial catalysts and it is possible to operate reasonably close to conditions of practical interest.

List of Symbols

C_i	Concentration of species i on surface sites
C_l	Concentration of free surface sites
$k_{\pm i}$	Reaction velocity constant for ith step

K_i	Equilibrium constant for ith step
K_{eq}	Overall equilibrium constant
p_i	Partial pressure of ith species
$V_+^{1,2,3,\ldots,n}$	Overall forward velocity
$V_-^{1,2,3,\ldots,n}$	Overall reverse velocity
V	Net overall reaction velocity per unit mass of catalyst
$v_{\pm i}$	Reaction velocity in forward and reverse direction for ith step
v_i	Stoichiometric number of ith step
v_r	Stoichiometric number common to all steps in a reaction path or stoichiometric number of rate controlling step.

References

Bard, G. (1974). "Nonlinear Parameter Estimation." Academic, New York.
Calderbank, P. H. (1952). *J. Appl. Chem.* **2**, 482.
Carberry, J. J. (1976). "Chemical and Catalytic Reaction Engineering." McGraw-Hill, New York.
Froment, G. F. and Bischoff, K. B. (1979). "Chemical Reactor Analysis and Design." Wiley, New York.
Happel, J., (1978). *AIChE J.* **24**, 368.
Happel, J., and Csuha, R. S. (1971a). *AIChE J.* **17**, 927.
Happel, J., and Csuha, R. S. (1971b). *J. Catal.* **20**, 132.
Happel, J., and Mezaki, R. (1970). *Chem. Eng. Sci.* **25**, 1952.
Happel, J., and Mezaki, R. (1974). *Chem. Eng. Sci.* **29**, 1300.
Happel, J., and Sellers, P. H. (1983). *Adv. Catal.* **32**, 273.
Happel, J., Hnatow, M. A., and Mezaki R. (1970). *Adv. Chem. Ser.* **97**, 92.
Happel, J., Kamholz, K., Walsh, D., and Strangio, V. (1973). *I&EC Fundam.* **12**, 263.
Happel, J., Suzuki, I., Kokayeff, P., and Fthenakis, V. (1980). *J. Catal.* **65**, 59.
Happel, J., Cheh, H. Y., Otarod, M., Ozawa, S., Severdia, A. J., Yoshida, T., and Fthenakis, V. (1982). *J. Catal.* **75**, 314.
Happel, J., Hnatow, M. A., Bajars, L., Otarod, M., and Lee, A. L. (1983). *Proc. Int. Gas Res. Conf.* Chicago, Illinois.
Happel, J., Walter, E., and Lecourtier, Y. (1986). *Ind. Eng. Chem. Fundam.* In press.
Hill, C. J. (1977). "An Introduction to Chemical Engineering Kinetics and Reactor Design." Wiley, New York.
Horiuti, J. (1973). *In* "The Use of Tracers in Heterogeneous Catalysis." J. Happel and M. A. Hnatow, eds.), *Ann. N.Y. Acad. Sci.* **213**, 5.
Hougen, O. A., and Watson, K. M. (1947). "Chemical Process Principles," Vol. III. Wiley, New York.
Mezaki, R., Chao, J. C., and Happel, J. (1980). *Chem. Eng. Sci.* **35**, 2361.
Neiman, M. B., and Gal, D. (1971). "The Kinetic Isotope Method and Its Applications." Elsevier, Amsterdam.
Otarod, M., Ozawa, S., Yin, F., Chew, M., Cheh, H. Y., and Happel, J. (1984). *J. Catal.* **84**, 156 (1983); erratum **89**, 564.
Park, S. W., and Himmelblau, P. M. (1982). *Chem. Eng. J.* **25**, 163.
Slin'ko, M. G. (1980). *Kinet. Katal.* **21**, 71.

Smith, J. M. (1981). "Chemical Engineering Kinetics," 3rd Ed. McGraw-Hill, New York.
Snagovskii, Y. S., and Ostrovskii, J. M. (1976). "Modeling the Kinetics of Heterogeneous Catalytic Processes" (in Russian), Khimia, Moscow.
Strangio, V. (1973). Doctoral thesis, New York University.
Temkin, M. I. (1971). *Int. Chem. Eng.* **11**, 709.
Temkin, M. I. (1979). *Adv. Catal.* **28**, 173.
Van Meerten, R. Z. C., Vollenbrock, J. G., de Croon, M. H. J. M., van Niselrooy, PF. M. T., and Coenen, J. W. E. (1982). *Appl. Catal.* **3**, 29.
Weller, S. A. (1956). *AIChE J.* **20**, 59.
Yang, K. H., and Hougen, O. A. (1950). *Chem. Eng. Progr.* **46**, 146.

Index

A

Adsorption
  associative, 30
  dissociative, 30
Ammonia synthesis
  mechanism, 56, 114
  rate equations, 114

B

Butane dehydrogenation
  mechanism, 65

C

Carbon monoxide oxidation, 138
Catalyst
  composition change, 2
  definition, 1
  heterogeneous, 2
  intermediate, 2, 28–31
  kinetics, 2
  surface study, 20–26
Catalyst composition, variation, 78
Catalyst reactions
  electronic, 31
Catalytic substances, 31–35
  acid–base reactions, 31
  ambient conditions, 34
  metals, 32, 33
  oxidation–reduction reactions, 31
  oxides, 32, 33, 34
  reaction types, 31
  semiconductors, 34
  stability, 37
  sulfides, 32, 33, 34
  transition elements, 33
Chemical reaction engineering, 5
Chemical reactions, independence, 41–44, 48
  Gibb's rule of stoichiometry, 42
Chemical reactor analysis, 5
Chemisorbed complexes, 28–31
  associative adsorption, 30
  dissociative adsorption, 30
  examples, 29, 30
  formation rate, 29, 30
  physical adsorption, 29
  properties, 29
Compartment definition, 8
Complexes, chemisorbed, 28
Computation procedure, curve fitting, 145
Correlation matrix, 148

D

Data fitting, generalized, 140–149
Direct mechanisms
  definition, 45
  enumeration, 49
  mathematical aspects, 45
Distinguishability, 129–132
  definition, 130
  example, 130
  mathematics, 131–132

E

Elementary reaction, 35–40
  active centers, 35
  definition, 6
  elucidation, 35
  kinetics, 38
  modeling procedure, 39
  molecularity, 35
  site density, 37
  virtual mechanism, 35
Ethylene oxidation
  mechanism, 60, 62, 63
Exponential modeling, problems, 138, 139

G

Gauss–Newton method, 143
General mechanism
  determination, 47
  mathematical aspects, 47
Goodness-of-fit, statistical, 146

H

Horiuti methods, 55, 102, 105
Hydrogen electrode reaction mechanisms, 54

I

Identifiability, 124–129
  definition, 123
  example, 126
  mathematics, 125
  testing, 125

Independence of reactions, 2, 41, 44
Intermediates, 3, 4, 7–9, 15
  chemisorbed, 2, 7
  specification, 28
Isobutane dehydrogenation
  kinetics, 176
  mechanism, 110
Isotopic studies
  chemisorption, 16
  exchange reactions, 17
  heterogeneous catalysis, 16
  isotopic kinetic effects, 19
  reaction path studies, 17
  reaction rates, 18
  relative rates, 17
Isotopic tracers, 3, 7, 10, 78–82
  mechanistic modeling, 3

K

Kinetics, 5
  LHHW models, 4, 10, 36, 174–176
  pseudo-monomolecular, 45
  rate equations, 174
Kinetics of multiple reactions, 183–187
  overall reaction space, 184
Kinetics of simple reactions, 175–183

L

Langmuir–Hinshelwood–Hougen–Watson (LHHW) rate equations, 174–176

M

Marquardt method, 143
Mass spectrometers, 15
Mathematical model, definition, 8
Matrix
  atom-versus-species, 43, 44, 60
  cycle-versus-step, 56, 58, 70
  step-versus-species, 50, 53, 54, 57, 62, 64, 65, 67
  variance–covariance, 147
Mechanism
  ammonia synthesis, 56
  butane dehydrogenation, 65
  cycle-free, 49

definition, 6, 44
description, 1, 6
direct, 45, 49
ethylene oxidation, 60, 62, 63
general, 47
Langmuir, 7
minimal, 45
Rideal, 7
sulfur dioxide oxidation, 53
synthesis gas methanation, 67
Mechanism modeling, 76–99
  assumed rate, 76
  kinetic data, 76
Mechanistic steps
  consideration, 36
  rate formulation, 37
Methanation of synthesis gas, 67, 153
  deuterium tracing, 163
  distinguishability, 159
  identifiability, 156
Minimal mechanism
  definition, 45
Model
  compartmental, 8, 86
  distinguishability, 9
  general, 82
  identifiability, 9
Modeling, superposition, 86–97
Model testing, 133–136
  ammonia synthesis, 135
  catenary, 133
  reaction steps, 133
  solutions, 139
Multiple-path reactions, 117

P

Parameter, determination, 123
Parameter set, admissible, 124
Potential factor, 10, 186

R

Rate equations, 173–190
  carbon monoxide methanation, 182, 185
  ethylene oxide production, 184
  isobutane dehydrogenation, 176
  LHHW models, 174, 175, 179
  power law models, 180
  reversible reactions, 175
  sulfur dioxide oxidation, 178
Reaction
  elementary, 6
  forward, 104
  reverse, 104
Reaction mechanism, 41–75
  atomic coefficients matrix, 43
  definition, 44
  dehydrogenation of butane, 65
  direct mechanisms, 49
  ethylene oxide formation, 62, 63
  independent reactions, 42
  matrix analysis, 47–49
  mechanism space, 46, 47
  multiple overall reactions, 61
  overall reaction, 63
  reaction space, 41
  reaction vector, 47
  simple, 52–60
  synthesis gas methanation, 67
Reaction modeling, 4, 5
Reactor
  backmixed, 12
  batch, 11
  bed, 11
  Bennett, 15
  Berty, 13
  Carberry, 12
  constant volume, 92–94
  continuous-flow stirred, 5, 11
  differential, 12
  flow, 11
  gradientless, 11, 86–92
  isothermal, 11
  plug-flow, 94–97
  recirculating, 13
  recycling, 14
  stirred, 12
Reversible process
  ultimate concentrations, 2
Routes, reaction, 173

S

Simultaneous reactions, 2
Site density
  calculation, 37

definition, 37
mechanism determination, 38
reaction rate, 37
State-space models, 125
Steady state
　definition, 77
　isotopic transfer, 4
　reaction conditions, 3
Steady-state hypothesis, 77
　characteristics, 77
　constant volume reactors, 78
Steady-state tracing, 100–116
　introduction, 100–102
　isobutene hydrogenation, 110
　multiple-path reactions, 117
　single-path reactions, 107
　step-velocity grouping, 102
　thermodynamics, 105
　transition state, 105
Stoichiometric equation
　definition, 1
Stoichiometric number, 55, 61, 102
Structural global identifiability, 124
Structural local identifiability, 125
Structural output distinguishability, 130
Sulfur dioxide oxidation
　mechanism, 53
　rate expressions, 178
　recirculating reactor, 118
　vanadium catalyst, 118
Superposition modeling, 86–92
　constant volume reactor, 86
　mathematics, 86
　plug-flow reactor, 94
　recirculating reactor, 89
Surface intermediates, 7
　characterization, 21
　experimental methods, 21–24
　nature, 20
Surface species, adsorbed, 2
Synthesis gas methanation
　mechanism, 67
Systems equations, 123

T

Temkin rate expressions, 104, 183
Terminal species
　definition, 5
　specification, 28
Tracer experimentation, 78–82
　catalyst impregnation, 81
　isotopic marking, 79
　kinetic information, 79
　labelling possibilities, 79
　reversibility, 80
　tracer incorporation, 81
　tracing variation, 79
Tracer modeling, mathematics, 82–85
Tracer studies, 149–162
　carbon monoxide oxidation, 149
　deuterium tracing, 163
　methanation, 153
　methanation distinguishability, 159
　methanation identifiability, 156
　multiple marked species, 162
Transfer function approach, identifiability, 125
Transient concentration, description, 123
Transient data, interpretation, 8
Transient testing, data fitting, 140
Transient tracer
　surface concentration, 4
　technique, 4, 7, 8
Transient tracing, 138–149
　data fitting, 140
　mathematics, 140–149
　problems, 139
　recirculating reactor, 138
　solutions, 139
　stirred-tank reactor, 138
Transition-state theory, 105

V

Variance–covariance matrix, 147
Virtual mechanism, definition, 35

RAYMOND H. FOGLER LIBRARY
DATE DUE

BOOKS ARE SUBJECT